Human Acid–Base Physiology

Human Acid–Base Physiology

A student text

Oliver Holmes
DSc, MSc, MB, BS, MRCS, LRCP, BA

Senior Lecturer in Physiology
University of Glasgow
Scotland

CHAPMAN & HALL MEDICAL

London · Glasgow · New York · Tokyo · Melbourne · Madras

Published by Chapman & Hall, 2-6 Boundary Row, London SE1 8HN

Chapman & Hall, 2-6 Boundary Row, London SE1 8HN, UK

Blackie Academic & Professional, Wester Cleddens Road, Bishopbriggs, Glasgow G64 2NZ, UK

Chapman & Hall Inc., 29 West 35th Street, New York NY 10001, USA

Chapman & Hall Japan, Thomson Publishing Japan, Hirakawacho Nemoto Building, 6F, 1-7-11 Hirakawa-cho, Chiyoda-ku, Tokyo 102, Japan

Chapman & Hall Australia, Thomas Nelson Australia, 102 Dodds Street, South Melbourne, Victoria 3205, Australia

Chapman & Hall India, R. Seshadri, 32 Second Main Road, CIT East, Madras 600 035, India

First edition 1993

© 1993 Oliver Holmes

Typeset in 10/12 pt Times by Thomson Press (India) Ltd., New Delhi
Printed in Great Britain by T. J. Press (Padstow) Ltd., Padstow, Cornwall

0 412 47610 X

A catalogue record for this book is available from the British Library

Library of Congress Cataloguing-in-Publication data

Holmes Oliver
    Human acid–Base physiology : a student text / Oliver Holmes.
        p.   cm.
    Includes bibliographical references and index.
    ISBN 0–412–47610–X
    1. Acid–Base equilibrium. 2. Acid–Base imbalances. I. Title.
QP90.7.H65 1993
612'.0 22—dc20   92–41400
                                                                    CIP

∞ Printed on permanent acid-free text paper, manufactured in accordance with the proposed ANSI/NISO Z 39.48-199X and ANSI Z 39.48-1984

Contents

Preface   xi
Units of measurement   xiii

1 FUNDAMENTALS   **1**

Acid–Base physiology   1
Water, the substrate of life   1
Water in the body   1
The dissociation of water in relation to acids and bases   2
Strong acids and weak acids   3
The law of mass action and the concentration of hydrogen
and hydroxide ions   3
Dissociation of water   4
Measurement of pH   7
Titration curves and buffers   7
Making a buffer solution   8
The Henderson-Hasselbalch equation   9
The chemical reactions accompanying buffering   10
The pK of a buffer pair and buffering efficacy   11
Buffer value   12
The importance of the CO_2–bicarbonate system   14
Kidney   16
Properties of gases: 17
    Partial pressure   17
    Avogadro's law   17
    Dalton's law of partial pressures   17
    The concentration of a constituent of a gas mixture   17
    Henry's law   18

Appendix 1   18
A.1 Ionic product of water   18
A.2 Definition and measurement of pH   18
A.3 Buffer value   19

2 INTRODUCTION TO ACID–BASE PHYSIOLOGY 21

The range of plasma pH in health and disease 21
Variations in ionic composition of the extracellular fluid
compatible with life 22
Comparison of hydrogen ions with other electrolytes 22
Enzymes and hydrogen ion concentration 22
Effects of disturbances of hydrogen ion concentration 23
Defence of hydrogen ion concentration 23
Buffering action of the body 24
Graphic representation of acid–base status 24
A graphic representation 25
Graphs for real fluids 27
Distilled water 27
Sodium bicarbonate solution 29
Perfect buffer 30
Real buffer solutions 30
Blood plasma 31
Blood 31

Appendix 2 32
A.1   The chemistry of carbon dioxide 32
A.2   Carbonate 32

3 DISORDERS OF ACID–BASE PHYSIOLOGY 33

3.1   Respiratory disorders 33
Respiratory acidosis (acute hypoventilation) 33
Compensation 36
Degrees of compensation 36
The time scale of compensation 37
The new blood line 37
Definition of acidaemia, acidosis etc. 38
Reinstatement of normal ventilation 40
Respiratory alkalosis 40

3.2   Metabolic disorders 41
Blood buffers 41
Metabolic acidosis 42
Infusion of hydrochloric acid 43
The role of bone in longer-term buffering 44
Compensation 44
Degree of compensation 45
Time scale 45
An advantage of acidaemia 46
Renal compensation 46
The final status 46

Cell membranes are permeable to blood gases but relatively
impermeable to ions 47
Treatment 47
Metabolic alkalosis 47
Potassium and acid–base physiology 48
The renal handling of electrolytes 49
Overview of hydrogen ion–potassium ion inter-relationships 51
Adverse effects of changes in extracellular potassium
concentration 51

3.3 Gastric function 52
The composition of gastric juice 52
Interactions between responses to disturbances of acid–base
physiology and of fluid volume such as vomiting 53
Vomiting of gastric contents 54
Outline of treatment 57

Appendix 3 57
A.1 Renal compensation for respiratory acidosis 57
A.2 Causes of metabolic acidosis 58
A.3 Renal handling of potassium 58

4 ASSESSMENT OF ACID–BASE STATUS 59

4.1 Standard bicarbonate, base excess and buffer base 59
Measurement of acid–base status 59
Indicators of the metabolic component 60
The standard bicarbonate 60
Base excess 62
Buffer base or total buffer base 63
The effect of changes in P_{CO_2} concentration on blood chemistry 65
Why are base excess and buffer base both used? 65
Specification of acid–base status 67

4.2 The Siggaard-Andersen nomogram for the parameters of
acid–base status 70
The effect of haemoglobin concentration 74
Review of methods of plotting acid–base status 77

Appendix 4 79
A.1 Measurement of total buffer base 79
A.2 $\log [CO_2]$ as a function of pH for a constant value of
$[HCO_3{}^-]$ 79

5 ACIDS AND BASES IN THE BODY 81

Sources of acid or alkali 81
Buffering of an acid load 82

Acid–base physiology in the context of body fluids 83
Composition of plasma 85
Total buffer base of plasma and of whole blood 86
The charge on one ion of protein buffer base 86
The blood as a buffer 87
The Henderson-Hasselbalch equation expressed in terms of
carbonic acid 88
The complex nature of blood and its acid–base implications 90
The 'chloride shift' 91
Blood in the body 92
Treatment of acid–base disorders 94
Intracellular hydrogen ion concentration 96
Comparison of extra- and intracellular hydrogen ion
concentration in disorders of acid–base physiology 96
Homeostasis 96

Appendix 5 99
A.1   Carbon dioxide and carbonic acid 99

6   OXYGEN CARRIAGE, RESPIRATION AND
    ACID–BASE PHYSIOLOGY 101

6.1   Carriage by the blood of oxygen and carbon dioxide 101
The oxygen dissociation curve of the blood 101
Typical values for P_{CO_2} and oxygen content in arterial
and mixed venous blood 103
The Bohr effect 104
Acid–base balance and the carbon-dioxide dissociation curve of
the blood 106
Uptake and release in capillaries of carbon dioxide 108
Comparison of the oxygen and carbon dioxide dissociation
curves of blood 109
The respiratory quotient 110
The difference in pH of systemic arterial and venous blood 110
Erythrocytes swell as they traverse systemic capillaries 111
Carbon dioxide and respiratory acidosis 111

6.2   The respiratory system in acid–base physiology 114
The composition of atmospheric air 114
Inspired air and alveolar gas 114
The partial pressure of gases in alveolar gas and in systemic
arterial blood 115
The body regulates the partial pressure of carbon dioxide in the
systemic blood 115

So long as hypoxia is excluded, the body does not regulate the
partial pressure of oxygen in the systemic blood 116
The partial pressure of nitrogen 116
The see-saw of carbon dioxide and oxygen 116
In hyperventilation, the body is depleted of carbon dioxide
but is not loaded with extra oxygen 117
Blood oxygenation and disorders of acid–base physiology 119
The control of ventilation 120

Appendix 6 121
A.1 'Oxygen dissociation' curve of the blood 121
A.2 The pH of blood as it passes through a systemic capillary 121

7 **RENAL ASPECTS OF ACID–BASE PHYSIOLOGY** 123
The mechanism of the renal secretion of hydrogen ions 123
Hydrogen ion secretion along the nephron 128
Excretion of acid 128
Phosphate as a buffer of urine 129
Ammonia 129
  Ammonia formation in the body 130
  Ammonia and renal acid–base physiology 130
Titratable acid and total acid excretion 132
Renal failure 132
Diuretic therapy 132

Appendix 7 133
A.1 The bicarbonate–sodium co-transporter for movement from
    the renal tubular intracellular fluid to the renal interstitial
    fluid 133
A.2 The production of ammonia from glutamine 133

8 **SELF-TEST QUESTIONS AND ANSWERS** 135

LEARNING OBJECTIVES IN ACID–BASE PHYSIOLOGY 169

REFERENCES 177

GLOSSARY 179

UNITS AND ABBREVIATIONS 183

INDEX 185

Preface

This textbook of human acid–base physiology is intended for pre-clinical medical students and other university students studying physiology courses of equivalent standard. This is a self-contained basic physiological text, in which the science emerges from a consideration of the physiological aspects of clinical acid–base disorders and is developed from basic principles. In writing the book, I have tried to produce a logical and sequential text in which the relevant physical chemistry is reviewed (Chapter 1), the essentials of acid–base physiology are established (Chapters 2 and 3), the assessment of acid–base status is described (Chapter 4) and the subject of acid–base physiology is put into perspective in the wider context of the physiology of the whole organism (Chapters 5 to 7).

The main points developed in the text are gathered in tables. This provides the reader, as information is assimilated, with a précis which is subsequently useful for reviewing or revising. Key references are included, so that the inquisitive can read more widely, and there are frequent allusions to clinical relevance. Furthermore, the book has been kept to a size that can reasonably be read by a junior undergraduate student.

The text is fully illustrated with original line diagrams. The final chapter is a substantial section of self-assessment material (including structured exercises and multiple choice questions), enabling readers to monitor their own progress. A glossary of terms is provided to help readers to understand the jargon of the subject and to check definitions etc.

An informal style of writing has been adopted with the hope of conveying the enthusiasm that the lecturer can achieve in the lecture theatre. A draft of this book has been extensively used by university undergraduates studying acid–base physiology and its effectiveness as a teaching text has been established on the campus.

ACKNOWLEDGEMENTS

I wish to express my indebtedness to the following people who have helped me so much in writing this book:

For remorselessly pointing out ambiguities and inadequacies of explanations: my students.

For providing invaluable expert guidance: Dr R. F. Burton, Professor Sheila Jennett and Mr T. Malone, who spent many hours on the manuscript.

For drawing the diagrams: Mr I Ramsden of the Medical Illustration Department, University of Glasgow.

For support with publishing the book: the staff of Chapman & Hall, in particular Mr P. Remes, with whom it has been a great pleasure to work.

Remaining errors are my responsibility, and I would be most grateful to have them drawn to my attention.

Oliver Holmes, 1992.

Units of measurement

The concentration of a chemical **A** in solution is indicated by square brackets thus: [**A**]. The units of measurement are :

Mole, or 'gram molecule', is an amount of a chemical; it is the molecular weight in grams. Example: for water, one Mole is 18 grams (Atomic weights: $H = 1$, $O = 16$).

M (abbreviation for 'Molar') represents moles per litre, or molar.

mM represents millimoles per litre (there are 1000 millimoles in one mole).

nM represents nanomoles per litre (there are 1000,000,000 nanomoles, or 10^9 nanomoles, in one mole).

pH $= -\log[H^+]$, where the concentration $[H^+]$ is in moles per litre.

Equiv., an abbreviation for 'Equivalent', is a measurement of quantity of charge; equiv. only applies to ions. It is moles multiplied by valency.

mEquiv. represents milliequivalents (there are 1000 milliequivalents in one Equiv.).

Equiv. per litre or **mEquiv. per litre** are measurements of concentration of charge.

Osmole, a measurement of quantity of osmotically active particles, is measured in the same units as Mole.

Fundamentals

1

ACID–BASE PHYSIOLOGY

Acid–base physiology is about the concentration of hydrogen ions in the body fluids. This first chapter is a summary of the physico-chemical background to the subject.

WATER, THE SUBSTRATE OF LIFE

Life as we know it could not have evolved without water. All living things contain water. Water has many remarkable properties of biological importance. It is a liquid at ordinary temperatures found on the earth. Compared with other liquids it has a high specific heat; this tends to prevent large changes in temperature when heat is produced by chemical reactions in the cells. It has a high latent heat of evaporation, providing the basis for an efficient mechanism for heat loss by evaporation of sweat. It is a good solvent for ionic compounds, which are essential components of all living systems.

WATER IN THE BODY

About two-thirds of the weight of the human body is water. Except for adipose tissue, all tissues contain a high proportion of water. The water is distributed into distinct 'compartments', the main subdivision being into water inside cells (intracellular water) and water outside cells (extracellular water). There is approximately twice as much intracellular water as extracellular water. The solutes in these two compartments are quite different and it is convenient to refer to two distinct fluids differing in composition, the **intracellular fluid** and the **extracellular fluid**.

THE DISSOCIATION OF WATER IN RELATION TO ACIDS AND BASES

All the body fluids are aqueous solutions. In aqueous solutions, water is not only the solvent but also takes part in chemical reactions; this is because the water molecules show a very weak tendency to dissociate into hydrogen ions and hydroxide ions:

$$H_2O \rightleftharpoons H^+ + OH^-$$

water   hydrogen   hydroxide
molecule   ion   ion

This reaction is shown as reaction 1 in Table 1.1, where the various reactions and equations are collected for reference as we proceed. In pure water, hydrogen ions and hydroxide ions are present in equal concentrations. When other chemicals are present, one or other may predominate. An **acid** solution is one in which the concentration of hydrogen ions exceeds that of hydroxide ions. An **alkaline** solution is one in which the concentration of hydroxide ions exceeds that of the hydrogen ions and a **neutral** solution is one in which the concentrations of the two types of ion are equal. Similarly, an acid is a chemical which releases hydrogen ions; an alkali, one which releases hydroxide ions.

A hydrogen ion is a proton, i.e. a hydrogen atom from which the electron has been removed. Over the years there has been confusion over the definition

Table 1.1

$H_2O \rightleftharpoons H^+ + OH^-$ Water hydrogen hydroxide ion ion		reaction 1
$HA \rightleftharpoons H^+ + A^-$ Acid hydrogen conjugate ion base		reaction 2
$NH_4^+ \rightleftharpoons H^+ + NH_3$ Ammonium hydrogen ammonia ion ion		reaction 3
$A + B \rightleftharpoons C + D$		reaction 4
$\dfrac{[C] \times [D]}{[A] \times [B]} = K$ Equilibrium constant		equation 1
$[H^+] \times [OH^-] = Kw \quad 10^{-14}$ at 25 °C Ionic product of water		

of acids and bases (Robinson, 1962, gives an interesting review of the history of the subject). The modern definition is as follows. An **acid** is a proton donor and a **base** is a proton acceptor (Bates, 1966, Guggenheim, 1957). The definition applies to the behaviour of the chemical in a particular solution in the particular conditions (temperature etc.) being studied. Certain chemicals, notably peptides and proteins, accept protons in a solution of high proton concentration and donate protons in a solution of low proton concentration; such chemicals behave accordingly as an acid or as a base depending on the ambient proton concentration. Since water itself dissociates, chemical equilibrium in aqueous media involves acids and bases both as solutes and also as ions contributed by dissociation of the solvent.

The distinction between the meanings of the words alkali and base is a fine one. An alkali is a hydroxide ion donor whereas a base is a hydrogen ion acceptor. Thus NaOH is an alkali whereas the hydroxide ion OH$^-$ is a base.

When an acid dissociates in an aqueous solution, it yields a hydrogen ion and a base; this base is called the **conjugate base** of the acid in question.

$$HA = \quad H^+ \quad + A^- \quad \text{(Table 1.1)}$$

Acid   Hydrogen ion   Conjugate base

In this reaction, the acid is a molecule and the base is an anion. This is commonly the case but not always so. For instance, ammonium ions (acid) dissociate to yield hydrogen ions and ammonia (base), as shown in the reaction:

$$NH_4^+ \quad = \quad H^+ \quad + NH_3 \quad \text{(Table 1.1)}$$

ammonium   hydrogen ion   ammonia
ion

STRONG ACIDS AND WEAK ACIDS

An acid HA which completely dissociates releases more hydrogen ions than an acid which only partially dissociates. In the latter case, reaction 2 is reversible. The acid which completely dissociates is called a strong acid and the one which partially dissociates is a weak acid. The less the tendency to dissociate, the weaker the acid.

THE LAW OF MASS ACTION AND THE CONCENTRATION OF HYDROGEN AND HYDROXIDE IONS

There is a relationship between the concentrations of hydrogen and hydroxide ions in a solution, a relationship which follows from the rates at which chemicals react. Suppose that two chemicals A and B react together thus:

$$A + B = C + D$$

The rate at which the reaction proceeds from left to right clearly depends on the frequency at which the particles of chemical A meet with those of chemical B; this in turn depends on the concentrations of A and B. If, for instance, the concentration of B is kept constant but the concentration of A is doubled, then the frequency at which particle A meets particle B is doubled and the rate at which the reaction proceeds is doubled. This result is generalized in **the law of mass action**: the velocity of a reaction at constant temperature is directly proportional to the product of the concentrations of the reacting substances. The concentrations are expressed in gram-molecules per litre.

On applying the law of mass action to determine the condition of equilibrium in the case of a reversible reaction taking place in a homogeneous system at constant temperature, represented by the general reaction shown above, one finds that the velocity of forward reaction **vf** is proportional to the product of the concentrations of A and B, or vf = kf × [A] × [B], where kf is the velocity coefficient (constant of proportionality) and the concentration, indicated by square brackets, is in moles per litre. Similarly, the velocity **vr** with which the reverse reaction takes place is given by the expression vr = kr × [C] × [D] where **kr** is the velocity constant for the reverse reaction.

If a certain amount of chemical A is added to a solution containing chemical B and the reaction is allowed to proceed, initially, the rate of backward reaction is initially zero because the concentrations of C and D are both zero. So initially the reaction only occurs in the forward direction. As time proceeds, the concentrations of C and D build up and, because of this, the rate of backward reaction increases. The concentrations of A and B are meanwhile falling because the chemicals are reacting together. Eventually, the rate of backward reaction catches up and reaches the rate of forward reaction. From this stage on, the rate at which A and B react to produce C and D equals the rate at which C and D react to produce A and B. Although the reaction is continuing in both directions, there is no longer any change in concentrations of any of the reactants. This is called **equilibrium**.

At equilibrium, the rates of forward and reverse reaction are equal; vf = vr so that

$$kf \times [A] \times [B] = kb \times [C] \times [D]$$

$$\frac{[C] \times [D]}{[A] \times [B]} = \frac{kf}{kr} = K \qquad \text{(Table 1.1)}$$

This equation gives the condition for equilibrium at a constant temperature, **K** being known as the **equilibrium constant**.

DISSOCIATION OF WATER

Since water ionizes to a slight extent, it is possible to apply to the equilibrium reaction 1 the law of mass action to obtain the expression:

$$\frac{[H^+] \times [OH^-]}{[H_2O]} = K$$

For any dilute aqueous solution, most of the solution is un-ionized water. Changes in the concentration of the ions are so small, by comparison with the concentration of the un-ionized water molecules, that the latter concentration may be regarded, to a close approximation, as being constant (section A.1). A new constant **Kw** may be defined incorporating K and $[H_2O]$ thus;

$$Kw = [H_2O] \times K$$

Equation 2 then simplifies to

$$[H^+] \times [OH^-] = Kw \qquad \text{(Table 1.1)}$$

The concentration of hydrogen ions in water at 25°C as determined from conductivity measurements is 10^{-7} molar (Findlay, 1953 p. 329), and since the concentration of hydroxide ions must be the same, the product of the concentrations is equal to 10^{-14}. This is the **ionic product** Kw of water.

$$Kw = 10^{-14}$$
$$[H^+] \times [OH^-] = 10^{-14}$$

For any neutral solution at 25°C, $[H^+]$ 10^{-7} molar, or 0.0000001 M.

This tiny concentration of hydrogen ions leads to difficulty in notation when expressing changes e.g. the difference between 0.00000001 and 0.0000001 M. To avoid all the zeros, the pH scale was introduced. The pH is the negative logarithm, to the base 10, of the hydrogen ion concentration in M (section A.2). Informally, pH is 'minus the exponent of the concentration'. The pH of a neutral solution is therefore 7.0 units at 25°C.

Another way of avoiding a string of zeros in expressing hydrogen ion concentration is to use nM (nanomole/litre) instead of M. The relationship between hydrogen ion concentration in nM and pH is shown graphically in Figure 1.1. On the nM scale, for a neutral solution at 25°C, the concentration of hydrogen ions is 100 nM. The dissociation of water increases as the temperature rises and, at 37°C, the concentration of hydrogen ions is 157 nM (Findlay 1953, p. 329).

A change of one unit in pH means a tenfold change in hydrogen ion concentration. Thus a solution of pH = 6.0 has ten times the concentration of hydrogen ions as a neutral solution; one with a pH = 8.0 has one-tenth the concentration of hydrogen ions.

A molar solution of a strong acid, strong enough to be completely dissociated, would have a hydrogen ion concentration of 1 M and, since $10^0 = 1$, it has a pH of zero. A 10 molar solution of acid has a pH of less than zero. For a strong alkali, a 1 molar solution, if completely dissociated, would have a pH of 14 and a 10 molar solution would have a pH over 14.

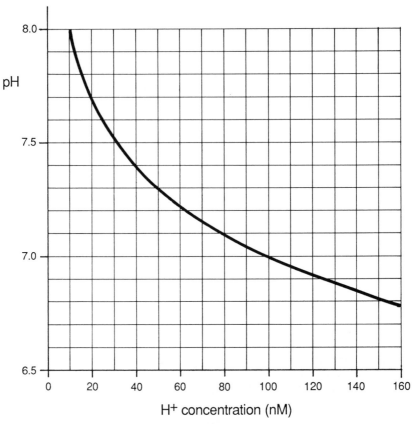

Figure 1.1. The relationship between hydrogen ion concentration (in nanomoles per litre) on the x-axis and pH on the y-axis. The graph allows ready conversion between the two scales. For instance, a pH of 7.1 corresponds to a $[H^+]$ of 80 nM. A ten-fold change in $[H^+]$ corresponds to a change in pH value of one. So to calculate the $[H^+]$ at pH = 6.1, we merely multiply 80 nM by 10 to give 800 nM (remembering that as the pH goes down, the $[H^+]$ goes up). In this way the graph can be used to convert between any pH and any $[H^+]$.

These calculations are hypothetical, since even the strongest acids and alkalis will not be completely dissociated at these high concentrations. However, the calculations indicate that the pH scale runs from negative values to values in excess of 14; it is not a scale which runs from zero to 14, as is erroneously stated in some elementary texts. The actual range of possible pH values, from measurements on very strong acids and alkalis, is from approximately −0.3 to 14.5.

MEASUREMENT OF pH

It has already been mentioned that in pure water the concentration of ions can be measured by conductivity measurements, but this method cannot be used when ions other than hydrogen and hydroxide ions are also present because the other ions would conduct also. A method which does not suffer from this disadvantage is the use of a pH electrode. In essence, this consists of an insulated tube sealed at the end by an extremely thin layer of specially prepared glass. This glass has the property that it is permeable only to hydrogen ions. The tube is filled with a solution with a known pH and is dipped into the test solution whose pH is to be measured.

The voltage between the two solutions is measured and this indicates the pH of the solution being tested. The pH meter scale is calibrated directly in units of pH.

TITRATION CURVES AND BUFFERS

A titration curve for a particular solution shows the extent of the change in pH for progressive additions of a strong acid or alkali; it is plotted thus. A litre of the solution is taken and its pH measured before adding any acid or alkali. Then, from a burette, a measured amount of strong acid at a known concentration is added, the solution stirred and the new pH measured. This is repeated many times and then a graph plotted of the pH as a function of amount of strong acid added. On the alkaline side, the procedure is repeated but using a strong alkali instead of the strong acid.

The relationship so plotted depends on the chemical nature of the solution at the start of the experiment. Figure 1.2 shows titration curves for two fluids. For pure water, addition of acid increases the concentration of hydrogen ions by a value which is equal to the amount of acid added divided by the total volume of fluid. For additions of alkali, the concentration of hydroxide ions increases by a value similarly calculated. The relationship for water in Figure 1.2 is curvilinear because pH is a logarithmic measure.

For many solutions, including the body fluids whose behaviour comprises the subject of acid–base physiology, additions of acid or alkali cause smaller changes in hydrogen ion concentration than those which occur when pure water is used. Such solutions are said to **buffer** the hydrogen ion concentration. The titration curve for a buffer solution which buffers most effectively at a pH of 7 is also shown in Figure 1.2. This buffering is a desirable feature in body fluids, since large fluctuations in pH would disrupt the harmonious interactions of enzymes needed for efficiency in biochemical reactions in the body. The salient features of a buffer solution are described next.

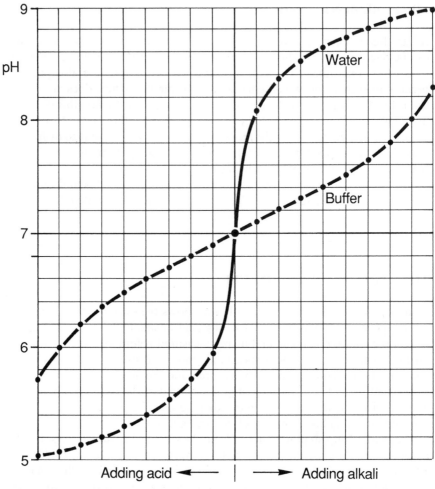

Figure 1.2. Titration curves for pure water and for a buffer solution which buffers most strongly at pH = 7.0. The points represents the pH values measured with a pH meter when measured amounts of acid or alkali are added. The steep slope in the region of pH = 7.0 of the relationship for water reflects the lack of buffering power of water. For the buffer, the smaller the slope, the more powerful the buffer.

MAKING A BUFFER SOLUTION

Every buffer consists of a buffer pair. In most cases the components are a weak acid and its salt. To make a buffer solution, a weak acid in solution is mixed with a salt of its conjugate base (e.g. acetic acid and sodium acetate). The resultant chemical reactions in solution are shown in Table 1.2. Since the acid is weak, it only partially dissociates, as shown in reaction 1 of Table 1.2. The

Table 1.2   Buffers

A buffer consists of a buffer pair; it is a mixture of a weak acid and its salt.

$$\text{Weak acid HA} \rightleftharpoons H^+ + A^- \qquad \text{reaction 1}$$
$$\text{Salt NaA} \rightarrow Na^+ + A^- \qquad \text{reaction 2}$$

Buffer pair consists of buffer acid HA
and buffer base A^-

$$\frac{[H^+] \times [A^-]}{[HA]} = K \Rightarrow [H^+] = K\frac{[HA]}{[A^-]} \qquad \text{equation 1}$$

Henderson-Hasselbalch equation

$$pH = pK + \log\frac{[A^-]}{[HA]} \qquad \text{equation 2}$$

pK is the pH when $[HA] = [A^-]$

Buffering is most effective for pH values within one pH unit on either side of the pK.

salt in solution completely dissociates as shown in reaction 2 of Table 1.2. In the solution are: weak acid, hydrogen ions, conjugate base and sodium ions. It is the weak acid and its conjugate base which provide the basis for the buffering action of the solution. This is the buffer pair previously mentioned. Of the buffer pair, HA is called the **buffer acid** and A^- is called the **buffer base**.

THE HENDERSON-HASSELBALCH EQUATION

We apply the law of mass action to the equilibrium reaction 1 in Table 1.2. In just the same way as, in Table 1.1, reaction 4 led to equation 1, here, in Table 1.2, reaction 1 leads to the expression:

$$\frac{[H^+] \times [A^-]}{[HA]} = K$$

Then

$$[H^+] = K \times \frac{[HA]}{[A^-]} \qquad \text{(Table 1.2)}$$

The aim is to find an expression for pH; to do this we take the negative logarithm of each side to give:

$$-\log[H^+] = -\log K - \log\frac{[HA]}{[A^-]}$$

Consistent with the definition of pH as the negative logarithm of the hydrogen ion concentration, we define the pK as the negative logarithm of the dissociation constant K; this leads to the expression:

$$pH = pK + \log\frac{[A^-]}{[HA]} \qquad \text{(Table 1.2)}$$

This is the Henderson-Hasselbalch equation; it plays a prominent role in acid–base physiology.

THE CHEMICAL REACTIONS ACCOMPANYING BUFFERING

When a strong acid such as hydrochloric acid (HCl) is added to a buffer pair, reaction 1 shown in Table 1.3 occurs. The hydrochloric acid, being a strong acid, completely dissociates to yield hydrogen ions and chloride ions. The conjugate base A^- of the buffer to which the acid is added has its charge balanced by a cation such as sodium. These four ions are on the left of reaction 1. The rise in hydrogen ion concentration due to addition of the hydrochloric acid drives the reaction to the right, by a mass action effect. The conjugate base of the buffer associates with the added hydrogen ions to yield buffer acid. The sodium ions and chloride ions remain free.

Since Na^+ and Cl^- appear on both sides of the equation, the reaction can be written in the abbreviated form shown as reaction 2 in Table 1.3. This serves to clarify the overall result; most of the added hydrogen ions react with buffer base. The change in pH produced by the added acid is less than if no buffer

Table 1.3 The chemical reactions accompanying the buffering of a strong acid or alkali

---

A. Add a strong acid to a buffer

$$H^+ + Cl^- \;+\; A^- \;+\; Na^+ \;\rightarrow\; HA \;+\; Na^+ + Cl^-$$

| Added acid | Buffer base | | | Buffer acid | | reaction 1 |

Abbreviated to

$$H^+ \;+\; A^- \;\rightarrow\; HA$$

| Added hydrogen ions | Buffer base | Buffer acid |

In a closed system, $[A^-]$ decreases, $[HA]$ increases.

B. Add a strong alkali to a buffer

$$OH^- \;+\; HA \;\rightarrow\; H_2O + A^-$$

| Added hydroxide ions | Buffer acid | | Buffer base | reaction 3 |

In a closed system, $[HA]$ decreases, $[A^-]$ increases

---

were present. The concentration of buffer base $[A^-]$ decreases whereas that of the buffer acid $[HA]$ increases.

When a strong alkali is added to a buffer solution, it is the weak acid which provides the buffering power, as shown in reaction 3 of Table 1. In summary, added acid is buffered by buffer base with the production of buffer acid and added alkali is buffered by buffer acid with the production of buffer base.

These events have been considered as they occur in a fluid in a closed flask, from which there is no escape and where there is no source of either of the components HA or A^- of the buffer system. This is called a 'closed system'.

THE pK OF A BUFFER PAIR AND BUFFERING EFFICACY

The efficacy of a buffer varies at different pH values. For the buffer acid and buffer base, the descriptor 'conjugate' is used to indicate that both share common chemical groupings. The Henderson-Hasselbalch equation written in these terms is then

$$pH = pK + \log([\text{conjugate base}]/[\text{conjugate acid}])$$

The pK value depends on the chemical nature of the particular buffer acid being studied. The Henderson-Hasselbalch equation indicates that, when the concentration of conjugate base equals that of the conjugate acid, the pH of the solution equals pK (since log(1) is zero). **The pK is the pH at which $[HA] = [A^-]$** This is reproduced in Table 1.2, where information on the general behaviour of buffers is collected.

As explained in the previous subsection, added acid reacts with buffer base to yield buffer acid; addition of acid thus decreases the numerator and decreases the denominator of the ratio $[A^-]/[HA]$. This ratio changes least if numerator and denominator start at equal values. To give an example, suppose that $[H^-]$ and $[HA]$ are initially both set at ten units. Addition of one unit of acid decreases $[A^-]$ by one and increases $[HA]$ by one; the ratio changes from 10/10 to 9/11, i.e. from 1 to about 0.82; this is a change of 12%. Let us suppose now that initially $[A^-]$ equals two units and $[HA] = 18$ units. Addition of the same quantity of acid as before changes the ratio from 2/18 to 1/19, i.e. from 0.1 to .052, a change of 47%.

This example can be generalized. Whenever acid or alkali is added, there is a change in the concentration of both components of the buffer pair, the one increasing and the other decreasing by virtually the same amount. If one component of the buffer pair is present at a much lower concentration than the other component, then the percentage change of concentration of the component at low concentration will be enormous. This greatly alters the ratio of concentrations of components of the buffer pair and hence the hydrogen ion concentration. In this situation, buffering is weak. Conversely, when both components are at approximately the same concentrations, then the percentage changes produced by addition of acid or alkali are relatively small and buffering is strong. This illustrates the general rule that **the buffer pair has greatest buffering power when the pH equals the pK**. A buffer pair is usually considered to be a useful buffer within one pH unit on either side of its pK. At pH values further than this from the pK, the concentration of one or other component of the buffer pair is so small that insignificant buffering power is provided.

This variation in buffering power with pH is illustrated in Figure 1.2. The pK of the buffer pair in the buffer solution was 7.0 units. For pH values in the range 6.0 to 8.0, addition of acid or alkali produces much smaller changes in pH in the buffer solution than in water. Beyond this range, both solutions show similar changes in pH as more and more acid or alkali is added.

BUFFER VALUE

This is a measure of the buffering power of a solution. It is defined in terms of the slope of the titration curve at a particular pH. It is the amount of strong acid or alkali (in gram-molecules per litre) producing a change of one pH unit (Van Slyke, 1922). As noted above, the buffer value is greatest when the pH is equal to the pK of a buffer solution.

The buffer value of a 1 mM solution of any buffer at the midpoint of the titration curve is 0.57 mmole/litre/pH unit (Cohen and Kassirer, 1982 p. 10). Section A.3 explains this.

Three titration curves are shown in Figure 1.3A. The central regions are near to being flat, indicating a high buffer value. In this central region, the curves all have the same slope, illustrating the result of the previous paragraph. On either side of the pK, the slope becomes progressively steeper, indicating reduced buffer values at pH values far from the pK. The titration curve for the buffer solution in Figure 1.3B shows that, at even greater divergences from the pK values, there is a point of inflection on either side, labelled x in the figure, beyond which the titration curves bend out again. This is a reflection of the fact that the pH scale is logarithmic. The graph for water is also shown on Figure 1.3B, where the logarithmic effect is shown uncontaminated by buffering. Equal increments of alkali or acid produce equal changes in $[H^+]$ but on a logarithmic scale this results in a very non-linear appearance. The graphs in Figure 1.3A only show the behaviour of the buffers between the points marked x on Figure 1.3B.

The labelling of Figure 1.3A reinforces a feature already considered in Table 1.3. On the acid side of the pK, the buffer acid is at a high concentration compared with its conjugate base. This reflects the fact that, when acid is added to a buffer pair, hydrogen ions combine with the conjugate base to form buffer acid; the concentration of conjugate acid is high and that of conjugate base is low. Conversely on the alkaline side of the pK, the conjugate base is more concentrated than its conjugate acid.

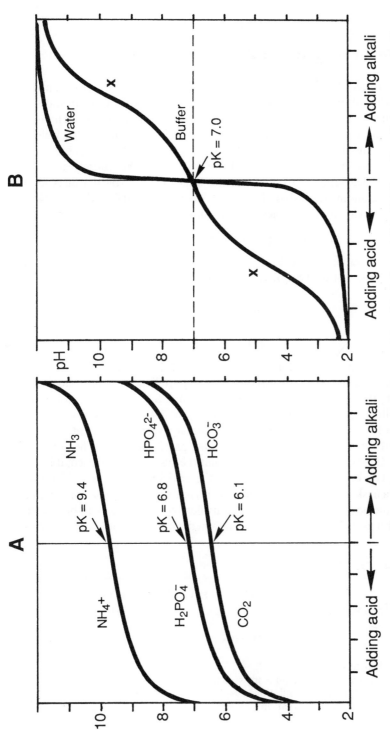

Figure 1.3. A. Titration curves for the three buffer pairs NH_4^+/NH_3, $H_2PO_4^-/HPO_4^{2-}$ and CO_2/HCO_3^-. For all three curves, buffer acid and buffer base are initially present in equal molecular concentrations and the pH is shown on the vertical line below the letter A; the pH here is equal to the pK of the buffer pair. Points corresponding to the addition of acid are shown to the left and of alkali to the right. On each curve, to the left of the pK value, the concentration of conjugate acid exceeds that of conjugate base whereas to the right, there is more conjugate base. This is symbolized by labelling with the conjugate acid on the left and conjugate base on the right. B. See text.

THE IMPORTANCE OF THE CO$_2$–BICARBONATE SYSTEM

As the subject of acid–base physiology is developed, it will become apparent that carbon dioxide and bicarbonate play a crucial role. Carbon dioxide is not itself an acid but in aqueous solution it reacts with water to yield hydrogen ions (Table 1.4 reaction 1) and such a chemical is an acid. It also yields bicarbonate and this is the corresponding (conjugate) base. The CO$_2$–bicarbonate system is thus a buffer pair in which we will refer, rather loosely, to CO$_2$ as the buffer acid and bicarbonate as the buffer base. Typical values for the concentrations in arterial blood are: $[CO_2] = 1.2$ mM and $[HCO_3^-] = 24$ mM.

The Henderson-Hasselbalch equation for this buffer pair is:

$$pH = pK + \log([HCO_3^-]/[CO_2]) \quad \text{(Table 1.4)}$$

and the pK value is 6.1 at 37°C.

Substitution of the typical values for $[HCO_3^-]$ and $[CO_2]$ gives:

$$pH = 6.1 + \log\frac{24}{1.2}$$

$$pH = 7.4$$

This is a typical value for the pH of extracellular fluid.

At first sight, one might discount the importance of carbon dioxide and bicarbonate as buffers of the extracellular fluid. Since the pK is 6.1 and the extracellular fluid is normally at a pH of about 7.4, the pK is more than one pH unit from body pH and, as we have already seen, this indicates that as a chemical buffer in a closed system, this buffer pair is inefficient. The importance of the CO$_2$-bicarbonate pair *in vivo* is that the system is not closed; it is called 'open'. The body is not isolated from the environment as is the case for a fluid in a flask. Both members of the buffer pair can be excreted from or retained within the body. CO$_2$, which is the equivalent of buffer acid, can be blown off by the lungs by hyperventilation or retained within the body by hypoventila-

Table 1.4 CO$_2$–HCO$_3^-$ system

$$CO_2 + H_2O \rightleftharpoons H^+ + HCO_3^-$$

Typical concentration 1.2 mM 24 mM reaction 1

$$pH = pK + \log\frac{[HCO_3^-]}{[CO_2]}, \quad pK = 6.1$$

CO$_2$ (equivalent to the buffer acid)
    is controlled by the respiratory system
    Blown off by hyperventilation
    Retained by hypoventilation
HCO$_3^-$ (the buffer base)
    is controlled by renal mechanisms

tion. In this way, the body can vary the concentration of this buffer acid as a compensatory mechanism for variations of hydrogen ion concentration.

As an example of respiration as a contributor to acid–base homeostasis, we compare the effects of adding strong acid to closed and open systems. For simplicity, extracellular fluid is considered. For the closed system, extracellular fluid is studied in a glass vessel and exchange of gases with the air is prevented. In this situation, almost all of the strong acid is buffered by bicarbonate. If 12 mM of strong acid is added to one litre of extracellular fluid, the bicarbonate concentration falls from its starting value of 24 mM to 12 mM; the $[CO_2]$ rises from its starting value of 1.2 mM to 13.2 mM. After addition of the acid, the pH is calculated thus:

$$pH = 6.1 + \log \frac{[HCO_3^-]}{[CO_2]}$$

$$pH = 6.1 + \log \frac{12}{13.2}$$

$$pH = 6.06$$

If the procedure is repeated except that carbon dioxide is free to interchange with a gas mixture containing CO_2 at a partial pressure of 40 mmHg, then in the Henderson-Hasselbalch equation, with addition of the acid, the concentration of bicarbonate changes as before but that of CO_2 remains constant:

$$pH = 6.1 + \log \frac{12}{1.2}$$

$$pH = 7.1$$

As will be explained later, the respiratory system contributes more than mere maintenance of a constant $[CO_2]$. When the pH is low as in the present example, hyperventilation contributes to the physiological response; by changing the P_{CO_2} to a subnormal value, this lowers $[CO_2]$ and thereby raises the pH to above 7.1. Such is the efficiency of the body in its response to disorders of acid–base physiology.

Not only is the concentration of carbon dioxide under physiological control, so also is the concentration of the other component of the buffer pair, the bicarbonate. This is controlled in the kidney, which may increase or decrease the excretion of bicarbonate as components of physiological control mechanisms. Since both components of the buffer pair are under physiological control, it is possible for the CO_2-bicarbonate system to act as a perfect buffer and completely to restore the pH of the internal environment to normal. Examples of such contributions will appear repeatedly as the subject of acid–base physiology unfolds.

KIDNEY

The kidney plays an important role as the organ for adjusting the plasma bicarbonate concentration. Bicarbonate is readily filtered in the glomerulus and is reabsorbed in the tubules; this is shown schematically in Figure 1.4A. Normally, almost all the filtered bicarbonate is reabsorbed. To lower the plasma $[HCO_3^-]$, the kidney reabsorbs less of the filtered load; bicarbonate is lost from the body. To raise the plasma $[HCO_3^-]$, the kidney reabsorbs all the filtered HCO_3^-.

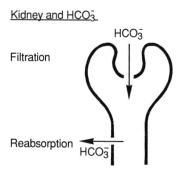

Kidney and HCO_3^-

Filtration

HCO_3^-

Reabsorption HCO_3^-

To lower plasma bicarbonate concentration, the kidney reabsorbs less of the filtered bicarbonate; this results in an increase in bicarbonate excretion and loss of bicarbonate from the plasma.

To raise plasma bicarbonate concentration, the kidney reabsorbs all filtered bicarbonate and generates 'new' bicarbonate from CO_2.

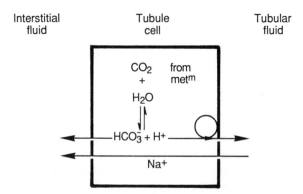

Interstitial fluid	Tubule cell	Tubular fluid

CO_2 from met^m
+
H_2O

$HCO_3^- + H^+$

Na^+

Figure 1.4. A. Diagrammatic representation of the glomerulus (the cup-shaped structure) leading to the tubule. B. The square represents a renal tubular cell.

If tubular reabsorption does not meet physiological requirements, a renal tubular mechanism for generating bicarbonate is called into play; this mechanism is shown in outline in Figure 1.4B. Carbon dioxide from metabolism reacts with water to yield hydrogen ions, which are actively pumped into the tubular fluid rendering the urine acid, and bicarbonate ions, which diffuse into the renal interstitial fluid and hence into the general extracellular fluid of the body. The tubules continually add bicarbonate to the body by this mechanism.

These movements of hydrogen and bicarbonate ions in opposite directions across the cell membrane of the tubular cells are of necessity accompanied by the movement of other charged particles to ensure that the net transfer of charge is zero. The principal contribution to provide this balance is by sodium ions, which move from tubular fluid to interstitial fluid as shown in Figure 1.4.

These renal mechanisms are described in more detail in Chapter 7. They are components of control mechanisms, providing the means for lowering or raising the plasma bicarbonate concentration as a compensation for disturbances of the hydrogen ion concentration, as described in Chapter 3.

PROPERTIES OF GASES

Partial pressure

The pressure which any one gas exerts, whether it is alone or mixed with other gases, is the 'partial pressure' of the gas. The partial pressure depends only on the number of moles of the gas in a given volume at a given temperature. The partial pressure of a gas is usually measured in mmHg, although other units are sometimes encountered.

Avogadro's law

The volume occupied by one gram-molecule of a gas is the same for all gases under the same conditions of temperature and pressure.

It is observed by experiment that one mole of a gas in a volume of 22.4 litres exerts a pressure of one atmosphere (760 mmHg) at $0°C$.

Dalton's law of partial pressures

When two or more gases are mixed, the total pressure is equal to the sum of the partial pressures of the constituents.

The concentration of a constituent of a gas mixture

This may be measured as the fraction of gas by volume, i.e. the ratio of the partial pressure of the constituent to the total pressure. Another unit is the percentage, which is the fraction of gas by volume multiplied by 100. By Avogadro's law, equal volumes of gas at the same temperature and pressure contain equal numbers of molecules and consequently the fraction of gas by volume is also the fraction of the gas by moles.

Henry's law

The amount of gas dissolved in physical solution in a liquid at a given temperature is directly proportional to the partial pressure of the gas. The constant of proportionality is the **solubility coefficient** of the gas.

APPENDIX 1

A.1 Ionic product of water

Since the concentration of the ions is very small, for any dilute aqueous solution, most of the solution is un-ionized water. The concentration of the un-ionized water molecules may be regarded to a close approximation as being 100%. This converted to molarity is the number of grams per litre divided by the molecular weight of water molecules. One litre contains 1000 g of water and the molecular weight of water is 18; the concentration of water is $1000/18 = 55.6$ moles per litre.

A.2 Definition and measurement of pH

The pH scale, as originally introduced by Sorensen, was defined as in this book as $pH = -\log[H^+]$ where the concentration $[H^+]$ was in units of moles per litre. Sorensen intended pH to be a measure of the effective hydrogen ion concentration but measurement with a hydrogen electrode gives pH as proportional to the logarithm of the activity (see below), not the concentration, of hydrogen ions. For dilute solutions, the activity and the concentration are nearly equal and, for the purposes of this book, Sorensen's original definition is used since it is adequately accurate for our purposes.

The pH scale is accepted on its merits because it is empirically indispensable. In practical situations, pH is deemed to be defined by the method of measurement. The pH value may be taken to be roughly $-\log[H^+]$. The higher the ionic strength of the solution, the greater the error in this assumption.

Activity. In using concentrations expressed in units such as moles per litre, one is assuming that ions and molecules do not interact with each other. In concentrated solutions, interactions may become highly significant and produce large modifications in the effective concentrations of solutes. For instance, a 1 M solution of glucose may give an osmotic pressure, as measured by experiment, of 20 atmospheres at $0°C$ instead of the theoretical 22.4 atmospheres.

To allow for this effect, the physicist substitutes an 'effective concentration' which is called the **activity** of a chemical in place of real concentration determined chemically. For dilute solutions such as extracellular fluid which are of primary concern in this book, concentrations and activities are sufficiently close for little error to be introduced by use of concentration. For intracellular fluid, however, the concentrations of organic compounds are relatively high and errors become significant; in this situation activities should be used for accurate analysis.

A.3 Buffer value

It was noted in the text that the buffer value of a 1 mM solution of any buffer at the midpoint of the titration curve is 0.57 mmole/litre/pH unit (Cohen and Kassirer, 1982 p. 10). At first sight, this is rather surprising; pH is a logarithmic scale so that a given pH change represents changes in $[H^+]$ depending on the absolute pH value. The reason, which arises from the Henderson-Hasselbalch equation, is this. If a unit amount of acid is added at pH close to pK, then almost all the added acid is taken up; the ratio [buffer base]/[buffer acid] is changed by removal of unit of buffer base and addition of a unit of buffer acid. For pH = pK + log(ratio), this gives the same change in ratio, and hence in log(ratio), independent of the pK. For the Henderson-Hasselbalch equation in its linear form (Table 1.2 equation 1), the change in ratio is multiplied by the K value (the dissociation constant) and hence varies directly with K. This gives the result that addition of a unit amount of acid to different buffer pairs each at the same concentration and each at their pK values gives the same change in pH, but changes in $[H^+]$ multiplied by K.

Introduction to acid–base physiology

2

THE RANGE OF PLASMA pH IN HEALTH AND DISEASE

The pH values of arterial plasma measured for all the individuals in a crowd of normal healthy people would fall mostly in the range 7.35 to 7.45, with an average of 7.4 (Robinson, 1962, p. 3). The plasma pH values measured on patients admitted to the metabolic ward of a hospital and suffering from untreated disorders of acid–base physiology would probably range between 7.1 and 7.7, this being the range compatible with life. For short periods of time, it is possible for the pH to go even further from normal particularly on the acid side, but these are useful limits to remember and are shown in Table 2.1.

Owing to the fact that the pH scale is logarithmic, this narrow range of pH does not indicate a very small range of concentration; in terms of hydrogen ion concentration expressed in nM, the corresponding values are shown in Table 2.1. A pH of 7.7 represents a hydrogen ion concentration of 20 nM and pH 7.1 represents 80 nM. So, in terms of absolute concentration, there is a four-fold range compatible with life. The tissues are relatively resistant to changes in hydrogen ion concentration.

Table 2.1

pH: range in health is 7.35 to 7.45, average is 7.4

Range in disorders of acid–base physiology compatible with life: 7.1 to 7.7

pH	7.7	7.4	7.1
$[H^+]$ nM	20	40	80
$[K^+]$ mM	2	4	8

$$[H^+] = K \frac{[CO_2]}{[HCO_3^-]}$$ equation 1

$$K = 0.8 \times 10^{-6}$$

For potassium, the relevant limits compatible with life are 2 and 8 mM; these values are also shown in Table 2.1. This similarity of numbers may act as a useful *aide-memoire*.

Bicarbonate and carbon dioxide are two chemicals of great importance in acid–base physiology and they too can deviate in concentration widely from normal before the change becomes life-threatening.

VARIATIONS IN IONIC COMPOSITION OF THE EXTRACELLULAR FLUID COMPATIBLE WITH LIFE

For sodium and chloride ions, a person would die long before the concentrations of these electrolytes rose to double or fell to half their normal value. This is associated with the fact that sodium and chloride are quantitatively the major electrolytes in the extracellular fluid and provide most of its osmotic pressure. It is the amount of sodium ions in the extracellular fluid which is the prime factor controlling extracellular fluid volume; maintenance of the concentration of sodium in this fluid is essential if the extracellular volume is to be maintained together with an adequate volume of circulating blood.

The osmotic effects of variation in concentration of a chemical such as bicarbonate are of less consequence than for sodium and chloride, because of the relative insignificance of the contribution of bicarbonate to osmolarity. Also, osmotic effects of changes in bicarbonate may be counterbalanced by an appropriate compensatory change in concentration of the electrolytes primarily responsible for osmotic pressure.

COMPARISON OF HYDROGEN IONS WITH OTHER ELECTROLYTES

The absolute concentration of hydrogen ions is measured in nM whereas all other electrolytes of interest in physiology have concentrations in the extracellular fluid measured in mM. There is a millionfold difference in the magnitudes of these two units; in absolute terms, hydrogen ion concentrations are far smaller than those of the other ions.

Hydrogen ions are also different from sodium, potassium and chloride in that they do not retain their identity in the body fluids but continually leave the hydrogen ion pool by combining with bases to form associated acids and then re-enter the pool when the acid dissociates again.

ENZYMES AND HYDROGEN ION CONCENTRATION

Enzymes consist of complex protein molecules, with many sites which attract and associate with hydrogen ions. Enzymic activity depends on the enzyme being in the correct state of ionization; if it is associated with an excess of hydrogen ions or has lost too many hydrogen ions, its enzymic activity is reduced or abolished. This is why enzymes operate optimally at a given pH. Disturbances of hydrogen ion concentration thus ultimately have as their adverse effect the interference with the normal harmonious interaction of the many thousands of enzymes on whose action cellular life depends. The effects at tissue or organ level are expressions of these enzymic disorders.

EFFECTS OF DISTURBANCES OF HYDROGEN ION CONCENTRATION

Deviation from normal pH usually occurs in association with other serious disturbances and it is difficult to specify the effects of altered pH alone. With a **high** hydrogen ion concentration, there is a widespread lowering of tone in smooth muscle; in vascular smooth muscle this results in a severe drop in arterial blood pressure, with circulatory collapse. When the high $[H^+]$ is prolonged, leeching of minerals from bones causes the bones to become weak mechanically, the condition of osteoporosis.

A **low** hydrogen ion concentration occurs as a result of hyperventilation, in which carbon dioxide is blown off in the lungs. The condition occurs in certain otherwise normal people who hyperventilate as a reaction to stress. A frequent result is a reduced concentration of ionized calcium in the extracellular fluid, the condition of hypocalcaemia. The effect is **hypocalcaemic tetany**, which consists of involuntary uncoordinated contractions of skeletal muscles and of bizarre subjective sensations and numbness. These symptoms and signs are due to hyperexcitability of peripheral nerve fibres, which fire action potentials spontaneously. The mechanism is that the reduced hydrogen ion concentration results in dissociation of complexes between plasma albumin and hydrogen ions, with a binding of plasma calcium to the albumin sites which are vacated by the hydrogen ions. Hence there is a fall in the concentration of ionized calcium in the extracellular fluid and this results in the nervous hyperexcitability.

In surgical practice, hyperventilation occurs if a very ill patient on life support including mechanical ventilation is inadvertently overventilated. The dangers of severe alkalosis caused by hyperventilation are related to the associated potassium depletion, the mechanism of which is described in Chapter 3. This hypokalaemia (low concentration of potassium in the blood) results in ventricular arrhythmia and death due to ventricular fibrillation.

DEFENCE OF HYDROGEN ION CONCENTRATION

The body has three lines of defence: buffering by the buffer systems of the body (blood, extracellular fluid, intracellular fluid and bone), respiratory compensation and renal compensation. In any given subject, one or more of these systems may be compromised. For instance, in anaemia, the buffering of the blood is defective; in a disorder which is primarily respiratory, the primary disorder removes the respiratory system from contributing any compensation; and in renal damage the contribution of the kidneys is reduced.

BUFFERING ACTION OF THE BODY

The first line of defence against a change in $[H^+]$ is provided by the buffer systems of the body. The buffers of the blood and extracellular fluid are immediately available whereas those of intracellular fluid take a matter of minutes to become effective. Because of its relatively low blood flow, bone, with an immense buffering power, requires hours or days to become available.

GRAPHIC REPRESENTATION OF ACID–BASE STATUS

The reversible reaction of carbon dioxide with water to yield hydrogen ions and bicarbonate ions is fundamental in acid–base physiology.

$$CO_2 + H_2O \rightleftharpoons H^+ + HCO_3^- \qquad \text{reaction 1}$$

The reaction may be more complicated than this, a point taken up in the appendix to this chapter and also in Chapter 8. For the present purposes, the simple form will suffice.

Applying the law of mass action (Chapter 1) to this equilibrium reaction, one obtains the expression

$$[H^+] = k_1 \frac{[CO_2] \times [H_2O]}{[HCO_3^-]}$$

Since in a dilute aqueous solution $[H_2O]$ is approximately constant, a new constant K is defined $K = k1 \times [H_2O]$
Then

$$[H^+] = K \frac{[CO_2]}{[HCO_3^-]} \qquad \text{equation 1}$$

At body temperature, the constant K can be shown to equal 0.8×10^{-6}. This is shown in Table 2.1 for reference.

The dimensions of K: From equation 1, $[CO_2^-]/[HCO_3^-]$ is a ratio and therefore dimensionless. Hence K has the same dimensions as $[H^+]$, i.e. the dimensions of concentration.

A GRAPHIC REPRESENTATION

If the values for any two of the three variables $[H^+]$, $[CO_2]$, $[HCO_3^-]$ are known, the third value is determined by equation 1 above. Any pair of the variables can be chosen for specifying concentrations of all three in an aqueous solution. It will prove useful in developing the concepts of acid–base physiology to indicate these concentrations on a graph with two of the variables as axes. Each choice of pairs of variables has advantages and disadvantages. The graphic representation chosen in this book is to plot $[HCO_3^-]$ as a function of PCO_2 or $[CO_2]$; the reasons for this choice will be discussed later. The axes to be used in plotting acid–base status are shown in Figure 2.1. The first objective is to investigate some features of this representation.

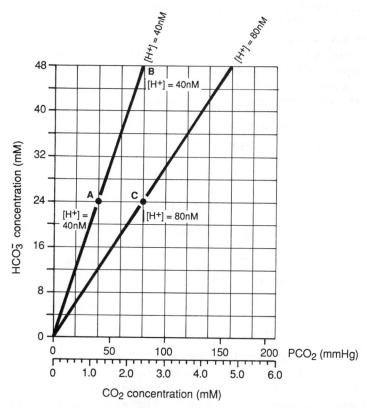

Figure 2.1. Bicarbonate concentration as a function of concentration of carbon dioxide in physical solution. The x-axis shows two equivalent calibration scales, in units either of PCO_2 in mmHg or of molecular concentration, in millimoles per litre. Each oblique line joins points at which the hydrogen ion concentration is constant, this being a consequence of the Henderson–Hasselbalch equation.

The x-axis. This is the concentration of carbon dioxide in physical solution. There are two numerical scales. By **Henry's law**, in a dilute solution, the concentration of a gas in physical solution is directly proportional to the partial pressure of the gas at a given temperature. The constant of proportionality is the solubility coefficient of the gas. For carbon dioxide, this gives the relation:

$$[CO_2] = \mathbf{a} \times P_{CO_2}$$

a is the solubility coefficient of CO_2 in water. With $[CO_2]$ measured in mM and P_{CO_2} measured in mmHg, **a** is 0.03 (at 37°C). It is as a consequence of this relationship that the x-axis has two equivalent scales of P_{CO_2} and $[CO_2]$. This will prove convenient since it allows immediate conversion between the two scales.

The y-axis. This is simply the bicarbonate concentration in mM. It is instructive to consider some particular points on the graph.

Construction of lines joining points of equal $[H^+]$ (also called 'isohydric contours' or iso-pH lines).

Point A. From equation 1 (Table 2.1) and the given value of K, $[H^+]$ in nM at the point **A** ($P_{CO_2} = 40$ mmHg, $[HCO_3{}^-] = 24$ mM) is calculated, these being typical values for human systemic arterial blood. For the calculation, the x-axis scales are used to convert 40 mmHg to $[CO_2]$ in mM; the answer is 1.2 mM. Substituting these values in equation 1 gives:

$$[H^+] = 0.8 \times 10^{-6} \times (1.2 \times 10^{-3})/(24 \times 10^{-3})$$
$$= 0.4 \times 10^{-7}\,M.$$

or

$$40 \times 10^{-9}\,M$$

or

$$40\,nM.$$

Point B. Now the hydrogen ion concentration is calculated for the case of both P_{CO_2} and $[HCO_3{}^-]$ being changed by the same factor. To take a specific example, doubling both $[CO_2]$ and $[HCO_3{}^-]$ gives the non-physiological point **B** ($P_{CO_2} = 80$ mmHg, $[HCO_3{}^-] = 48$ mM).

$$[H^+] = 0.8 \times 10^{-6} \times (2.4 \times 10^{-3})/(48 \times 10^{-3}),$$

i.e.

$$[H^+] = 40\,nM$$

This illustrates the general rule that changing P_{CO_2} and $[HCO_3{}^-]$ by the same factor results in no change in pH. In retrospect, the reader will recognize this as being self evident from equation 1. It follows that any point on a straight line through the origin and through point A has a $[H^+]$ value of 40 nM. This is shown in Figure 2.1 as the oblique labelling '$[H^+] = 40$ nM' at the top of the graph; this labelling applies to the corresponding oblique line.

Point C. For this, the P_{CO_2} is doubled but the $[HCO_3{}^-]$ is kept constant. The result, from equation 1, is that the hydrogen ion concentration is doubled.

So point C has $[H^+] = 80\,nM$. A second straight line joining points with this $[H^+]$ is constructed and labelled with this $[H^+]$.

In order to read off the $[H^+]$ for any particular point on the graph, a ruler is positioned to join the origin of the axes with the point in question and then read from a scale of $[H^+]$ on the top of the graph. In Figure 2.2, the x and y axes are the same as in Figure 2.1 and in addition, scales of $[H^+]$ and pH are provided around the top of the graph for this purpose.

Thus far, the equation relating $[H^+]$, $[HCO_3^-]$ and $[CO_2]$ has been investigated on the graphic representation of acid–base status; real fluids have not yet been considered. This is the next topic.

GRAPHS FOR REAL FLUIDS

For any fluid it is possible to investigate experimentally how the bicarbonate concentration increases as the P_{CO_2} with which it is in equilibrium is increased. The experimental procedure is as follows. The fluid of interest, be it a chemical solution made up in the laboratory or a sample of plasma from a patient, is equilibrated with a gas containing carbon dioxide at a known partial pressure. This gives the x coordinate to plot on the graph. The concentration of bicarbonate in the fluid is measured. This gives the y coordinate and defines one point on the graph. Next, the fluid is equilibrated with a gas containing a different partial pressure of carbon dioxide and the new concentration of bicarbonate is measured. This gives a second point on the chart. In this way, as many points are plotted as are needed to indicate the whole relationship, which is finally filled in as a curve.

For every fluid, a rise in P_{CO_2} will always cause the chemical reaction:

$$CO_2 + H_2O \rightleftharpoons H^+ + HCO_3^- \qquad \text{reaction 1}$$

to move to the right; there will therefore be a rise in $[HCO_3^-]$. This rise is in some instances very small, as will shortly emerge, but the $[HCO_3^-]$ can never fall with a rise in P_{CO_2}. The plot on the chart must be flat or sloping up; it can never slope downwards.

The first relationship to be considered is that between P_{CO_2} and bicarbonate concentration for water. Then fluids are considered which contain plasma constituents important in acid–base physiology.

DISTILLED WATER

The effects of changes in P_{CO_2} on $[HCO_3^-]$ in distilled water are illustrated by deriving two points from the relationship. One point is given for water equilibrated with a gas containing no CO_2. Water does not contain any bicarbonate; the coordinates of the point A are therefore (0,0) in Figure 2.2. A second point is determined as follows.

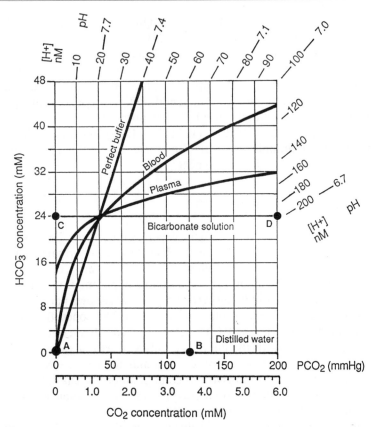

Figure 2.2. As for Figure 2.1, but more information displayed. The scale around the top shows hydrogen ion concentration both in units of nanomoles per litre and in units of pH. The relationships are shown for distilled water, 24 mM bicarbonate solution, plasma and blood. The perfect buffer does not exist in a closed system, but physiologically the extracellular fluid is an open system and physiological mechanisms are able in some situations to ensure perfect buffering, as shown by the line so labelled on this graph.

As the partial pressure of carbon dioxide in the gas equilibrating with the water is increased, every molecule of CO_2 which dissociates yields one H^+ ion and one HCO_3^- ion. If the tiny concentration of hydrogen ions contributed by the dissociation of water is neglected, then: $[H^+] = [HCO_3^-]$. Incorporation of this constraint into equation 1 indicates that water equilibrated with CO_2 at different partial pressures obeys the expression:

$$[HCO_3^-]^2 = 0.8 \times 10^{-6}[CO_2]$$

The particular value for the P_{CO_2} of 120 mmHg is selected. Then $[CO_2] = 3.6 \times 10^{-3}$ M and

$$[HCO_3^-]^2 = 0.8 \times 10^{-6} \times 3.6 \times 10^{-3}.$$

It will soon emerge that it is sufficiently accurate to calculate to the nearest power of ten.

$$[HCO_3^-]^2 < 10^{-8}.$$
$$[HCO_3^-] < 10^{-4} M \text{ or } 1/10th \, mM$$

i.e. less than one-tenth mM. This cannot be distinguished from zero on the scale of the y-axis. (This is the justification for working to the nearest power of ten). The calculations give the point B (120, approx. 0) on the graph in Figure 2.2.

There are now two points on the relationship for distilled water, both indistinguishable, on the scale being used, from the x-axis. These considerations lead to the conclusion that the relationship for water between P_{CO_2} and bicarbonate concentration is indistinguishable from the x-axis, as shown in Figure 2.2.

SODIUM BICARBONATE SOLUTION

The next fluid for consideration is a solution of sodium bicarbonate; a concentration of $[HCO_3^-] = 24 \, mM$ is chosen since this is a typical concentration of bicarbonate for arterial blood plasma from a healthy person. As before, the point representing $P_{CO_2} = 0$ is plotted. $NaHCO_3$ is a salt and so completely dissociates in water.

$$NaHCO_3 \rightarrow Na^+ + HCO_3^-$$

24 mM $NaHCO_3$ in solution contains both $[Na^+]$ and $[HCO_3^-]$ at a concentration of 24 mM. This yields the point C (0,24) in Figure 2.2.

For a second point on the relationship, a high value for the x-coordinate, or P_{CO_2}, is chosen, just as for water. Although the $[HCO_3^-]$ can be calculated, the calculation is tedious and a more intuitive approach is adopted. The $[HCO_3^-]$ is estimated when the $P_{CO_2} = 200 \, mmHg$. Since the logic which follows is quite closely argued, it is set out in steps, the first of which is trivial.

Step 1. When the $P_{CO_2} = 200 \, mmHg$, $[CO_2] = 6 \, mM$.

Step 2. Every molecule of CO_2 which reacts with water yields one ion of H^+ and one of HCO_3^-.

Step 3. $[HCO_3^-]$ must be at least 24 mM, since that was the concentration in the solution before exposing it to an increased P_{CO_2}.

Step 4. The point sought lies on the vertical line with x-coordinate 200 mmHg. The concentration of hydrogen ions along this line progressively decreases as the y-coordinate increases (note the crossing of hydrogen ion concentration contours as one ascends vertically anywhere on the chart). Since $[HCO_3^-]$ must be at least 24 mM, the y-coordinate of the sodium bicarbonate solution at this P_{CO_2} must be at least 24 mM. Reference to the scale showing the hydrogen ion concentration reveals that $[H^+]$ cannot be more than 200 nM.

Step 5. From step 2, it follows that the maximum possible additional bicarbonate yielded by carbon dioxide is also 200 nM or 0.0002 mM. The total $[HCO_3^-]$ is therefore not more than $(24\,mM + 0.0002\,mM) = 24.0002\,mM$.

On Figure 2.2, this cannot be distinguished from 24 mM. So, to a very close approximation, the point D (200 mmHg, 24 mM) is on the bicarbonate line. As for water, the relationship may be filled in for intermediate points and we arrive at the conclusion that, for sodium bicarbonate solution, the relationship is horizontal given the resolution of the scales chosen to display the graphs.

For both water and sodium bicarbonate, the fact that the relationship between $[HCO_3^-]$ and $[CO_2]$ is a straight line with zero slope reflects the fact that these solutions are totally lacking when it comes to buffering the hydrogen ions released when the P_{CO_2} increases. As the P_{CO_2} rises, the graph for each fluid crosses isohydric contours at the highest possible rate.

The principal point which this section illuminates is that, for sodium bicarbonate solution, increasing the P_{CO_2} results in an insignificant change of $[HCO_3^-]$ but a large change of $[H^+]$. Carbon dioxide and bicarbonate together thus lack significant buffering power for changes in P_{CO_2}. (It will be shown later that these chemicals act as a powerful buffer for acids and bases other than CO_2 and HCO_3^-.)

Conclusion: For any solution lacking buffering power, increasing the P_{CO_2} results in an increase in bicarbonate concentration which is negligible on the scale used to plot physiological concentrations. A plot of $[HCO_3^-]$ as a function of P_{CO_2} gives a relationship very close to a straight line with a slope of zero.

PERFECT BUFFER

The perfect buffer (i.e. a solution whose $[H^+]$ is constant no matter what acid or alkali is added) does not exist, but let it be assumed that an approximation to a perfect buffer can be produced. Suppose that such a solution holds the $[H^+]$ at 40 nM and that it contains 24 mM bicarbonate when the P_{CO_2} is 40 mmHg. These values are chosen since they represent the composition of arterial plasma from a normal person. The relationship of the perfect buffer with this composition is the straight line through point A and the origin (Figure 2.2).

REAL BUFFER SOLUTIONS

Since a solution with no buffering power gives a straight line with zero slope whereas a perfect buffer gives an oblique straight line through the origin of the axes, any real buffer solution must give an intermediate relationship. The relationship lies between the lines for perfect buffering and for zero buffering. For a solution with high buffering power, the graph will be close to the perfect buffer, i.e. a graph with a steep slope or gradient; for a solution with low buffering power, the graph will have a small gradient.

BLOOD PLASMA

This contains, amongst other chemicals, bicarbonate and the non-bicarbonate buffers plasma protein and phosphates. (Quantitatively, the plasma proteins are much more important than the phosphates in buffering.) These buffers have a limited buffering capacity so that the $[HCO_3{}^-]$ against P_{CO_2} curve for plasma has a small gradient. This is shown in Figure 2.2 as the curve labelled 'Plasma'. The graph passes through the point $[HCO_3{}^-] = 24\,mM$ and P_{CO_2} of 40 mmHg.

BLOOD

The erythrocytes contain haemoglobin at a high concentration and this contributes substantially to the buffering power of blood; the graph for whole blood has a greater gradient than that for plasma, as indicated in Figure 2.2. There is sufficient buffer reserve in blood to ensure that, as the P_{CO_2} is reduced towards zero, all the bicarbonate is converted to CO_2 which diffuses out of the blood and into the gas with which the blood is being equilibrated. On the chart this is indicated by the graph passing through the origin of the axes. For plasma, there is not enough of the necessary buffer to displace all the bicarbonate and so the graph intercepts the y-axis. This difference in behaviour is considered in more detail in a later chapter.

The relationship plotted for blood has several features of importance. Although some of these have already been given, they are restated here together for convenience.

The relationship plotted for blood:

a) starts from the origin of the axes;
b) passes through the 'physiological' point for arterial blood, i.e $P_{CO_2} = 40\,mmHg$, $[HCO_3{}^-] = 24\,mM$;
c) is a curve which is always rising; in this respect its behaviour must be carefully distinguished from that of the oxygen dissociation curve, which for high partial pressures of oxygen flattens out towards a line with zero slope (Chapter 6).

APPENDIX 2

A.1 The chemistry of carbon dioxide

When carbon dioxide is released by metabolism, two chemical reactions are available to it:

$$CO_2 + H_2O \rightleftharpoons H_2CO_3 \rightleftharpoons H^+ + HCO_3^- \qquad \text{reaction 1}$$

This is called 'hydration' of carbon dioxide.

$$CO_2 + OH^- \rightleftharpoons HCO_3^- \qquad \text{reaction 2}$$

This is called 'hydroxylation' of carbon dioxide; the OH^- is derived from the dissociation of water.

It is not known whether CO_2 produced by metabolism in the body is processed mainly by reaction 1 or mainly by reaction 2 (Valtin, 1983, p. 201). Either way, the final products are H^+ and HCO_3^- and, at equilibrium, the concentrations of CO_2 etc. are given by the Henderson Hasselbalch equation. In order to simplify the presentation in the main text, reaction 1 is used in its abbreviated form:

$$CO_2 + H_2O \rightleftharpoons H^+ + HCO_3^-$$

and the Henderson-Hasselbalch equation arrived at for this chemical reaction. This simplified approach makes for ease of understanding and introduces no error.

A.2 Carbonate

A chemist concerned with the reactions of bicarbonate would need to consider the chemical reaction:

$$2NaHCO_3 \rightarrow Na_2CO_3 + H_2O + CO_2$$

or the equivalent reaction:

$$HCO_3^- \rightarrow H^+ + CO_3^{2-}.$$

The production of carbonate only occurs at very low partial pressures of carbon dioxide, far lower than those occurring physiologically. In all circumstances of physiological interest, this dissociation of bicarbonate to carbonate and hydrogen ions is quantitatively negligible and may be ignored.

Disorders of acid–base physiology

3

Chapters 1 and 2 have provided the necessary background to consider disturbances of acid–base physiology and the body's responses, which forms the subject matter of this chapter.

SECTION 3.1 RESPIRATORY DISORDERS

Carbon dioxide is a respiratory gas and in aqueous solution is essentially an acid. Disorders of respiration therefore have serious acid–base consequences. When aeration of the blood in the pulmonary capillaries is impaired, accumulation of carbon dioxide leads directly to an acidification of the extracellular fluid; this is called **respiratory acidosis**. Conversely in circumstances when aeration of the blood in the pulmonary capillaries is excessive, carbon dioxide is washed out of the body, leading to the condition of **respiratory alkalosis**.

RESPIRATORY ACIDOSIS (ACUTE HYPOVENTILATION)

'Hypoventilation' means a condition in which pulmonary ventilation is reduced and is defined as a rise in the alveolar P_{CO_2} to a value in excess of 40 mmHg in a person breathing air. It occurs if the respiratory centres are depressed, as in certain conditions of drug abuse. It also occurs when there is an obstruction to the air passages, narrowing of the bronchioles by disease, bronchoconstriction, e.g. an attack of asthma, or by inhaling a foreign object. A dramatic example of the last-mentioned condition is afforded by a child who breathes in whilst holding a tin whistle in its mouth. The whistle is inhaled, sticks in the trachea and partially obstructs air flow through the trachea. This impairs aeration throughout the lungs. With a foreign object lodged in its trachea, the child will be making frantic respiratory movements, but adequate aeration of the blood is prevented by the obstruction to free air flow. The alveolar P_{CO_2} rises and, since the lungs are failing as a pump, this condition is properly described as hypoventilation. As this example demonstrates, 'hypoventilation' does not necessarily mean a depression of the movements of respiration, which is a common misconception amongst students.

Although the example of a foreign object lodged in the trachea is unusual, it is a useful illustration of the account which follows because the primary disorder starts suddenly, is maintained at a reasonably constant level and then may cease as suddenly as it started when the obstruction is removed.

The rise in the alveolar P_{CO_2} results immediately in an equivalent rise in the arterial P_{CO_2}. As a result, the reaction

$$CO_2 + H_2O \rightleftharpoons H^+ + HCO_3^-$$

is driven to the right with a consequent rise in the bicarbonate concentration and in the hydrogen ion concentration. It is useful to plot changes in acid–base

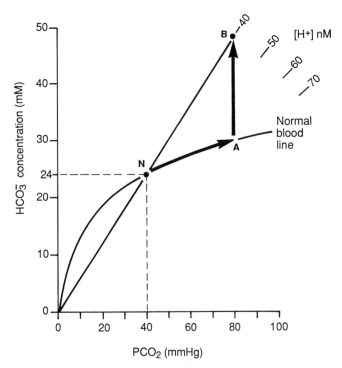

Figure 3.1. Bicarbonate concentration as a function of P_{CO_2}. The normal blood line is shown and the oblique line representing a normal extracellular hydrogen ion concentration. Point N is a typical point representing normal acid–base status. Arrow NA indicates the change accompanying uncompensated respiratory acidosis and arrow AB indicates renal compensation.

status on the graph of bicarbonate concentration as a function of P_{CO_2}, as described in Chapter 2. Figure 3.1 illustrates such a plot. The relationship labelled 'blood' in Figure 2.2 of Chapter 2 is reproduced in Figure 3.1 and labelled 'normal blood line'. The reason for this nomenclature is that, as we shall see, the blood in certain patients with disorders of acid–base physiology gives a graph which is displaced from the normal position.

In a patient in whom there is a sudden rise in the alveolar partial pressure of carbon dioxide, the change in composition of the patient's arterial blood is represented in Figure 3.1 as a move from the normal, represented by point N, along the normal blood line to point A. The increase in the x coordinate represents the rise in the P_{CO_2} of the patient's arterial blood. The increase in the y coordinate represents the rise in bicarbonate concentration; the move to a new isohydric contour reflects the increase in hydrogen ion concentration. These changes in composition of the arterial blood are summarized in Table 3.1 Column A.

The original disorder is a rise in arterial P_{CO_2} as shown in Table 3.1. The immediate results, due to mass action effects in the blood, are that the $[HCO_3{}^-]$ and $[H^+]$ rise. From the point of view of acid–base physiology, the rise in hydrogen ion concentration has been buffered to some extent chemically by the blood buffers. The time scale is the same as that of retention of CO_2, which starts instantaneously and is complete within a quarter of an hour. The condition is called **acute respiratory acidosis**. The descriptor 'acute' refers to the fact that these effects occur immediately, accompanying the accumulation in the body of carbon dioxide. At this stage there is as yet no physiological compensation for the rise in $[H^+]$ and hence the descriptor 'uncompensated'. The $[H^+]$ has risen and the kidneys can compensate for this. Although the renal physio-

Table 3.1 Hypoventilation

	A Uncompensated	B Renal compensation (bicarbonate retention)
P_{CO_2} mmHg	High 80	High 80
$[HCO_3{}^-]$ mM	High 29	High 48
$[H^+]$ nM	High 67	Normal 40
	Acute uncompensated respiratory acidosis	Compensated respiratory acidosis

logical mechanisms are called up immediately, it takes longer for their effects to become manifest. Such more prolonged effects are called 'chronic'.

COMPENSATION

As a result of the increased acidity of the blood, renal compensatory mechanisms come into play. The overall result is that the kidney returns to the body (in the renal venous blood) more bicarbonate than that entering it (in the renal arterial blood). The plasma bicarbonate concentration, elevated already by the buffering reactions, rises still further as a result of the renal mechanism. The change in acid–base status is indicated by a move vertically upwards, from A to B in Figure 3.1. Compensation does not mean the restoration of normal blood biochemistry; it means compensation for the change in hydrogen ion concentration. In the example of compensated respiratory acidosis, the pH is restored to normal but the $[CO_2]$ is high and the $[HCO_3{}^-]$ is pushed higher as a result both of the biochemistry accompanying buffering and the physiological compensation.

To recapitulate briefly: the basic defect is increased P_{CO_2}. For whatever reason, the subject cannot ventilate more and so cannot reduce the P_{CO_2}. Instead, the kidney adds bicarbonate to the blood as part of the physiological mechanism of minimizing changes in hydrogen ion concentration. The details of the renal mechanisms for this response are not fully understood but one contributing factor is that the raised P_{CO_2} in the renal tubular cells drives the reaction

$$CO_2 + H_2O \rightleftharpoons H^+ + HCO_3{}^-$$

to the right, releasing more hydrogen ions for transfer into the tubular fluid by the hydrogen ion pump mechanism (Chapter 7). This is a simple mass action effect.

The blood chemistry in compensated respiratory acidosis is summarized in Table 3.1 column B. At the risk of being repetitive, an important principle needs to be reiterated. In acid–base physiology, 'compensation' does not mean the restoration of normal blood biochemistry; it means compensation to bring the originally abnormal hydrogen ion concentration back towards normal. The pH is restored but the $[CO_2]$ and $[HCO_3{}^-]$ are both far from their normal values. Compensation is physiological buffering.

DEGREES OF COMPENSATION

As shown in Figure 3.2, the compensation for the changed hydrogen ion concentration may be partial, in which case the pH is partially restored but is still low; it may be complete, in which case the pH is normal; or over-compensation may occur, with the pH rising above normal (section A.1).

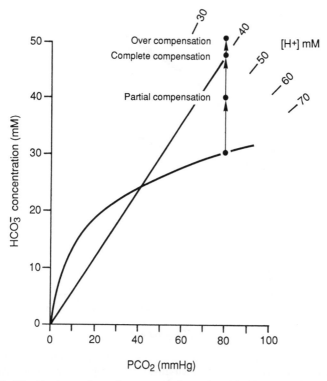

Figure 3.2. The degrees of renal compensation of respiratory acidosis. In partial compensation the hydrogen ion concentration is partially restored towards normal, in complete compensation the hydrogen ion concentration is fully restored and in over-compensation, the renal retention of bicarbonate has more than compensated for the rise in P_{CO_2}.

THE TIME SCALE OF COMPENSATION

The renal mechanisms come into play immediately the disturbance of acid–base physiology starts. However, it takes a relatively long period for the renal mechanisms to come to maximum efficiency and for the renal retention of bicarbonate to build up the concentration throughout the body. In respiratory acidosis, it takes around three days for the subject to settle to the compensated point. This is where the subject then stays so long as the same degree of hypo-ventilation persists.

THE NEW BLOOD LINE

When sufficient time has elapsed for the subject to have established a new steady state, the composition of the blood has changed radically. The kidney

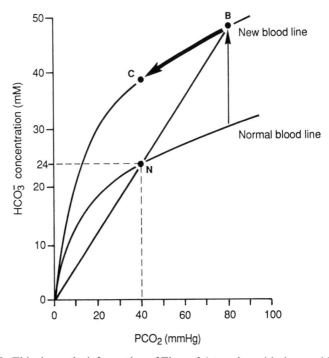

Figure 3.3. This shows the information of Figure 3.1 together with the new blood line of a subject who has completely compensated for the respiratory acidosis. Arrow BC indicates the change in blood chemistry if hypoventilation is suddenly corrected and the respiratory system is able to excrete the previously-accumulated CO_2.

has retained bicarbonate with the consequence that the blood contains a higher concentration of bicarbonate than that produced by the CO_2 retention alone. Arterial blood taken from a patient with complete compensation of hydrogen ion concentration yields a CO_2-bicarbonate relationship shown as the 'new blood line' in Figure 3.3. The bicarbonate retention by the kidneys has moved the subject to a new blood line, which is at a higher level than before.

So long as the hypoventilation persists, the patient remains in an abnormal steady state. The blood chemistry is represented by point B in Figure 3.4.

DEFINITION OF ACIDAEMIA, ACIDOSIS ETC.

Most authorities use the term 'acidaemia' to mean a low blood pH and 'alkalaemia' to mean a high blood pH (Table 3.2). 'Acidosis' means that the extracellular fluid contains excess of unwanted acid but there is not necessarily a low plasma pH; the pH may have been returned to normal values because

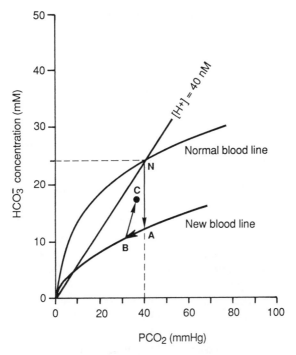

Figure 3.4. The changes in blood chemistry occurring in metabolic acidosis. Arrow NA indicates the blood buffering of the acid load, arrow AB indicates the respiratory compensation for the acidaemia and arrow BC indicates the renal contribution to compensation, with retention of bicarbonate.

Table 3.2   Definitions

Acidaemia: a low blood pH
Alkalaemia: a high blood pH
Acidosis: an excess of unwanted acid in the body; the pH may be normal
Alkalosis: an excess of unwanted alkali in the body, the pH may be normal

of compensatory mechanisms. Correspondingly, 'alkalosis' means that there is an excess of unwanted alkali in the extracellular fluid although the pH may be normal.

This nomenclature is now applied to the sequence of events in hypoventilation. Within half an hour, the subject moves to a point which is described as 'acute or uncompensated respiratory acidosis'. There is acidaemia. Over the course of the ensuing days, renal compensation occurs and the patient's blood is represented by a point which is described as 'compensated respiratory acidosis'. Since the pH has returned to normal, there is no longer acidaemia; the condition is properly described as acidosis because there is still an excess of unwanted CO_2 in the body.

REINSTATEMENT OF NORMAL VENTILATION

Suppose now that, after three days, the obstruction to ventilation is removed; normal respiratory exchange is resumed and over the course of minutes, the excess CO_2 is washed out of the body. In Figure 3.3, the point representing the composition of the subject's blood moves down along the new blood line until the P_{CO_2} has fallen to around normal values (point C). The bicarbonate concentration is still high, reflecting the bicarbonate retention by the kidneys.

The renal compensation is now the opposite, with the kidney losing bicarbonate. As with the original compensation, it takes a few days for renal mechanisms to have their effects in full and for normal blood biochemistry to be restored. During this time, the acid–base status gradually moves along the line from point C to point N in Figure 3.3.

This account has been of the stages in the response to hypoventilation which is of sudden onset and then maintained at a constant level. In other conditions, exemplified by a patient whose ventilation becomes progressively but slowly compromized by widespread chronic lung disease, the two phases of respiratory acidosis overlap temporally. It is then not possible to disentangle the uncompensated phase from the compensated phase. The point representing the blood chemistry moves slowly but obliquely between N, the point representing normal acid–base status, to point B, representing the contributions both of blood buffering and of renal compensation. In this chapter, hypoventilation of acute onset was considered precisely because it allows appreciation of responses with differing time scales.

RESPIRATORY ALKALOSIS

This is due to overbreathing. It occurs in the 'hyperventilation syndrome'; susceptible subjects hyperventilate, perhaps in response to stress, for minutes, days and even weeks. It also occurs in healthy people at high altitude, where ventilation is stimulated by hypoxia. In surgical practice, hyperventilation occurs in patients being overventilated by a mechanical ventilator. In this situation, CO_2 is washed out of the body and respiratory alkalosis supervenes. Because of the dangers of this situation, the arterial blood gas tensions should be frequently checked in such patients and the P_{CO_2} should not be allowed to fall below 30 mmHg.

The buffering which occurs in respiratory alkalosis and the changes in renal handling of bicarbonate are the converse of those in respiratory acidosis. It is left as an exercise for the reader to follow through these changes. An important additional factor, to be considered in detail later in this chapter, is a fall in plasma potassium concentration accompanying alkalosis. The greatest danger to life of alkalosis is the associated hypokalaemia; this results in arrhythmias of the ventricles of the heart and death caused by ventricular fibrillation.

SECTION 3.2 METABOLIC DISORDERS

Disorders of acid–base physiology which are not of respiratory origin are called 'metabolic' disorders. This nomenclature derives from the fact that such disorders result from abnormal metabolism. Metabolic disorders of metabolism may also be due to excessive intake of acid or alkali or to failure of renal function, when the tubular mechanisms for formation of acid or alkaline urine are impaired.

Respiratory disorders of acid–base physiology have the common feature that the primary abnormality is a deviation from normal of the amount of carbon dioxide in the body. Metabolic disorders are caused by a large variety of primary abnormalities resulting from an excess of non-respiratory acid or alkali in the body. It is useful to classify the causes thus:

1. too much acid or alkali taken into the body (by mouth, injection etc.);
2. excessive loss of acid or alkali through the gastrointestinal tract (e.g. vomiting, diarrhoea);
3. abnormal amounts of acid or alkali produced by abnormalities of metabolism;
4. depression of the kidney's ability to excrete acid (or alkali).

BLOOD BUFFERS

In order to understand metabolic disorders of acid–base physiology, blood buffers must be considered in a little more detail. It is useful to subdivide

Table 3.3 Blood buffers

CO_2–bicarbonate buffer pair
CO_2 corresponds to buffer acid
HCO_3^- buffer base

Non-bicarbonate buffers

HPr non-bicarbonate buffer acid
Pr^- non-bicarbonate buffer base
(Pr stands for 'protein')

the blood buffers into two groups. Firstly, there is the CO_2–bicarbonate buffer pair. Secondly, all the other blood buffers are grouped together as **non-bicarbonate buffers** (Table 3.3). Quantitatively the most important are haemoglobin and the plasma proteins. Other non-bicarbonate buffers, such as phosphate, contribute relatively little in quantitative terms in blood and plasma. As a first approximation, it is legitimate to ignore phosphate etc. and to consider all non-bicarbonate buffer to be protein buffer. As with all buffers, the protein buffer consists of a buffer pair, non-bicarbonate buffer acid written as HPr and non-bicarbonate buffer base written as Pr^- (Pr stands for 'protein').

METABOLIC ACIDOSIS

A brief classification of the causes of metabolic acidosis is given in section A.2.

Metabolic acidosis occurs in a healthy person who exercises maximally. The exercising muscles in this situation do not obtain the energy which they need from aerobic metabolism and so metabolize anaerobically. This results in the release of metabolites including lactic acid into the extracellular fluid.

An example of metabolic acidosis occurring as a result of disease is uncontrolled diabetes mellitus (Table 3.4). The tissues cannot metabolize glucose properly and instead excessively metabolize fat to yield so-called 'ketone bodies' which include aceto-acetic and beta-hydroxy-butyric acids. These are strong acids. Such acids are called 'non-volatile', or 'fixed' because they cannot be blown off in the lungs as can CO_2. Renal damage is another cause of metabolic acidosis; as will be explained later, a normal mixed diet yields a preponderance of acid when metabolized and the urine is normally acid. When renal function is impaired, the kidney fails to excrete this metabolically-derived acid and the acid accumulates in the body. In surgical patients, a common

Table 3.4 Metabolic disorders

Disorders of acid–base physiology not of respiratory origin are called metabolic disorders

Metabolic acidosis
Example 1. Diabetes mellitus
Accumulation of 'ketone bodies'

aceto-acetic acid ⎫
⎬ strong acids
β-OH-butyric acid ⎭

Non-volatile-unlike CO_2 they cannot be blown off in the lungs. Hence they are called 'fixed'

Example 2. Renal failure
Example 3. Circulatory failure

cause of severe metabolic acidosis is circulatory failure due to blood loss. With circulatory failure, there is inadequate perfusion of the tissues by blood, the tissues are hypoxic and so metabolize anaerobically. This mode of metabolism leads to a build-up of lactic acid and therefore metabolic acidosis.

When fixed acid accumulates in the body because of diabetes, the accumulation is in many cases very slow, over the course of days, weeks or months. Compensatory mechanisms are called up as the accumulation proceeds. So it is uncommon to meet patients with the sudden 'uncompensated' condition as is sometimes the case with respiratory disorders. However, to understand the different components of a metabolic disorder, it is instructive to follow the sequence of events in the artificial situation of a rapid injection of fixed acid, for instance of hydrochloric acid intravenously, into an anaesthetized animal. Such a situation has occurred in a human when, by tragic accident, a strong acid has been infused and even then, the phases of the responses of the body run into each other. It will nevertheless be easier if we consider the phases as if they occurred separately.

INFUSION OF HYDROCHLORIC ACID

In such a case, the first stage is uncompensated metabolic acidosis, in which excess H^+ is largely taken up by the blood buffers. The chemical reactions are shown in Table 3.5. Both types of buffer base, bicarbonate and non-bicarbonate, combine with hydrogen ions. For the bicarbonate system, the CO_2 yielded (reaction 2) is lost via the lungs. As a result of reaction 2 in Table 3.5, bicarbonate is removed from the blood to be excreted as CO_2 and the blood bicarbonate concentration falls.

This same information is shown in graphic form on the chart of acid–base status in Figure 3.4. Respiration is immediately stimulated by the rise in hydrogen ion concentration in the blood; its contribution is to be described shortly. In the very short term, since it takes a finite length of time for CO_2 to be washed out of the body, the $[CO_2]$ may, for the purposes of explanation, be regarded as initially unchanged and the fall in bicarbonate concentration,

Table 3.5 Infusion of HCl

First stage–uncompensated metabolic acidosis

$$H^+ + Pr^- \rightarrow HPr \qquad \text{reaction 1}$$

$$H^+ + HCO_3^- \rightarrow H_2O + CO_2 \qquad \text{reaction 2}$$

↓	↓	↓
From added acid	Buffer base in the blood	Excreted from the lungs

$[HCO_3^-]$ falls.

Table 3.6   Metabolic acidosis

	Uncompensated	Respiratory Compensation (Hyper-ventilation)	Renal Compensation (Bicarbonate reabsorption)
$[H^+]$ nM	High 80	Partially restored 72	Further restored 50
P_{CO_2} mmHg	Normal 40	Low 30	Low 35
$[HCO_3^-]$ mM	Low 12	Low 10	Partially restored 17

occurring in a few seconds, is a move vertically down, from point N, representing normal acid–base status, to the point A. There is a concomitant lowering of the blood pH, to a value which may be as low as pH 7.1, $[H^+] = 80$ nM.

In summary, as shown in the column labelled 'uncompensated' of Table 3.6, the hydrogen ion concentration has risen because of added acid, the concentration of CO_2 is unchanged and the bicarbonate concentration has fallen as a result of blood buffering. This is called 'uncompensated metabolic acidosis'. Since the plasma bicarbonate concentration falls with, initially, little alteration in arterial P_{CO_2}, it follows that the patient is operating on a new blood line. Arterial blood taken at this stage would yield the 'new blood line' in Figure 3.4. This contrasts with uncompensated respiratory acidosis described in the last section, where the point representing the subject's acid–base status moved to a point on the normal blood line.

THE ROLE OF BONE IN LONGER-TERM BUFFERING

The immediately available buffers are those of the blood and extracellular fluid. In the longer term, intracellular buffers and bone add their contribution; the buffers of intracellular fluid take a matter of minutes to become effective. The inorganic matrix of bone binds large amounts of sodium ions. In acidaemia, hydrogen ions displace and exchange with this sodium which provides a large reserve of buffering for acid. The uptake of acid by bone is relatively slow, taking hours or days.

COMPENSATION

For metabolic acid–base disorders, the body has two lines of defence. The first is respiratory compensation. In the case of metabolic acidosis, the fall in pH stimulates the peripheral chemoreceptors, leading to hyperventilation. The onset of hyperventilation is immediate; carbon dioxide is washed out of the body and the arterial P_{CO_2} falls below normal. The subject's acid–base status is represented as a move down along the new blood line. The hydrogen ion concentration is partially restored towards its normal value. This is shown as the arrow A B in Figure 3.4. This compensation is appropriately called 'Respiratory compensation for metabolic acidosis'.

The compensation involves a fall in plasma bicarbonate concentration below the fall caused by the initial chemical buffering of hydrogen ions. The fall in P_{CO_2} produced by the hyperventilation is proportionally greater than this accompanying fall in bicarbonate concentration. From the point of view of hydrogen ion concentration, the overall effect of the hyperventilation is to contribute to a restoration towards normal of the blood pH, with arrow A B indicating a move back towards a normal pH.

DEGREE OF COMPENSATION

As we have seen, in metabolic acidosis the raised hydrogen ion concentration stimulates respiration via the peripheral chemoreceptors. The resultant washing-out of CO_2 from the body lowers the arterial P_{CO_2} and thus reduces the central drive to respiration. A subject with metabolic acidosis hyperventilating at a constant level has arrived at a steady state, with respiration produced by the raised peripheral drive and the reduced central drive. This interaction was studied by Ypersele de Strihou and Frans (1970). In patients who were acidotic because of renal failure, the pattern of response of the respiratory control centres was such that, on average, each decrement of plasma bicarbonate of 1 mM evoked a 1.2 mmHg reduction in P_{CO_2} of the arterial blood; this corresponds to a rise in hydrogen ion concentration of just over 1 nM. The change in hydrogen ion concentration due to the fall in bicarbonate accompanying buffering of the fixed acid load is about halved by the respiratory compensation.

TIME SCALE

It takes up to a quarter of an hour for the hyperventilation to establish a new steady state, after which point B represents the acid–base status of the subject. The situation is shown in the column of Table 3.6 labelled 'respiratory compensation'. The respiratory compensation thus merges in time with the blood buffering. Access of the acid load to intracellular buffers takes minutes, during which time CO_2 is being washed out of the body. The contribution of bone to buffering is insignificant within the time span of respiratory compensation arriving at its steady state.

AN ADVANTAGE OF ACIDAEMIA

In cases of metabolic acidosis due to circulatory collapse, the acidaemia confers some physiological advantage in that it causes a shift to the right of the oxygen dissociation curve of the blood (Chapter 6). This assists in the unloading of oxygen from blood to the tissues and to a slight extent counteracts the tissue hypoperfusion.

RENAL COMPENSATION

The kidney plays its part and cooperates with the lungs in returning the blood pH towards normal. Renal effects run with a slower time course than respiratory effects. To counteract the rise in hydrogen ion concentration, the kidneys retain bicarbonate which is accompanied by an increase in the renal excretion of hydrogen ions and the urine is acid; this is an appropriate response since the kidney is thereby ridding the body of the acid load. The renal compensation carries the acid–base status along the arrow B C in Figure 3.4. The upward component of this arrow shows the retention of bicarbonate. As the kidney restores the plasma $[HCO_3{}^-]$ and hence the pH towards normal, the stimulation of respiration by the acidaemia is partially relieved. Ventilation returns towards normal and the degree of hyperventilation is reduced.

At this stage, the hydrogen ion concentration is only partially compensated; the partial pressure of carbon dioxide is low, as is the bicarbonate concentration. This is shown in the right panel of Table 3.6. In metabolic disorders, it is usual for compensation to be only partial, in contrast to disorders which are of respiratory origin.

THE FINAL STATUS

The final acid–base status of a subject with metabolic acidosis depends on the origin of the acidosis. With a bolus injection of acid, the renal mechanisms ultimately excrete all the acid load and return the blood chemistry to normal; the subject's status returns to point N in Figure 3.4. In a clinical condition such as diabetes, when fixed acid is being continually released into the body, the body settles to a steady state represented by point C a distance from N.

A point which was made earlier in the chapter should now be reiterated. In disease processes such as diabetes in which fixed acid is slowly released into the body fluids, the different components of the response to the acid load cannot be separated according to different time scales. The compensatory processes are gradually called up as the disease progresses. In Figure 3.4, the acid–base status point moves smoothly from N directly to C. The bicarbonate lost in the chemical buffering of the fixed acid is partially replaced by renal retention of bicarbonate.

CELL MEMBRANES ARE PERMEABLE TO BLOOD GASES BUT RELATIVELY IMPERMEABLE TO IONS

An example of the effects of the relative impermeability of membranes to HCO_3^- is afforded by the following clinical example. When there is an excess of non-respiratory acid in the body, there is a gradual acidification of the cerebrospinal fluid (CSF). The stimulus to the central chemoreceptors which stimulate the respiratory centres is the hydrogen ion concentration and, with acidification of the CSF, respiration is stimulated to the extent of producing hyperventilation. The treatment to correct the excess of acid is to administer bicarbonate intravenously. With this treatment, the plasma $[HCO_3^-]$ immediately increases, with the plasma hydrogen ion concentration being restored to normal. The blood-brain barrier is relatively impermeable to bicarbonate so that the CSF remains acid for some time; at this stage, the excess respiratory drive persists, the arterial Pco_2 remains low and the blood hydrogen ion concentration may fall below normal.

TREATMENT

The administration of bicarbonate is to be used with caution. The indiscriminate use of bicarbonate is particularly dangerous in resuscitation of patients with metabolic acidosis as a concomitant of hypovolaemic shock; lactic acid itself is innocuous and is readily removed by the liver as soon as the perfusion of the tissues is re-established. If administration of bicarbonate causes alkalosis and shifts the oxygen dissociation curve to the left, there is interference with oxygen unloading at the cellular level in tissues which are already hypoxic. Treatment of metabolic acidosis by bicarbonate therapy is reserved for situations in which partial correction of the pH is needed to restore cardiac function, which is depressed by acidaemia as described in Chapter 4.

METABOLIC ALKALOSIS

Metabolic alkalosis is produced iatrogenically in the case of a subject who habitually swallows baking powder (sodium bicarbonate) to allay the pain of duodenal ulcer. The responses of the body to a simple metabolic alkalosis are the converse of those for metabolic acidosis. It is left as an exercise for the reader to follow the changes occurring in this disturbance. Metabolic alkalosis also occurs clinically as a result of vomiting of gastric contents, but here the acid–base disturbance is only one of the problems with which the body must contend. This is explained in more detail in a later section.

POTASSIUM AND ACID–BASE PHYSIOLOGY

The details of the interaction of hydrogen and potassium ion concentration are complicated and poorly understood. An important general observation is that changes in the concentrations of potassium ions and of hydrogen ions in extracellular fluid usually move together, a rise in [H$^+$] accompanying a rise in [K$^+$] and a fall accompanying a fall. There are several mechanisms with this as a common feature. These mechanisms depend on the fact that there can be no net transfer of charge across a membrane.

One mechanism is shown in Figure 3.5A. When the primary disorder is

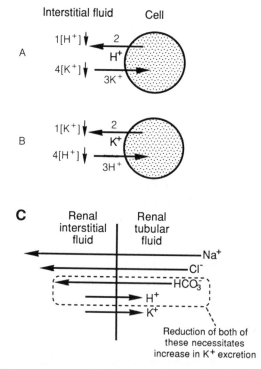

Figure 3.5. A. The movements of hydrogen and potassium ions between intra- and extracellular fluid compartments produced by a fall in hydrogen ion concentration of the extracellular fluid. B. Similarly for a fall in potassium ion concentration of the extracellular fluid. C. Movements of ions between the renal tubular fluid and the renal interstitial fluid: sodium, chloride and bicarbonate are reabsorbed, hydrogen and potassium ions are secreted.

alkalaemia, the fall in the hydrogen ion concentration in the interstitial fluid causes an outward movement of hydrogen ions from cells. This is because the plasma membrane is permeable, if only slightly, to hydrogen ions so a fall in external concentration gives rise to a concentration gradient in the outward direction and a consequent efflux of hydrogen ions. Preservation of electro-chemical neutrality requires a balancing flux of other ions across the membrane; some of this is contributed by potassium ions moving into the intracellular compartment, as shown in Figure 3.5A. This lowers the extracellular concentration of potassium.

The inverse sequence occurs if the primary abnormality is a fall in extra-cellular potassium concentration; the steps are shown in Figure 3.5B. In both A and B, a rise in concentration of one will entrain a rise in the other. For all cells other than those of the renal tubules, this mechanism, since it involves movements to and fro across the cell membrane, results only in movements of ions between compartments of the body without loss of total body content. The renal tubular cells are exceptional because movement of electrolytes from the cytoplasm of these cells into the tubular fluid leads to irretrievable loss of the electrolytes from the body.

THE RENAL HANDLING OF ELECTROLYTES

Sodium and chloride ions are readily filtered together with all other plasma electrolytes at the glomerulus. Typical plasma concentrations are 140 mM for sodium and 100 mM for chloride; the filtered load of sodium is therefore 40% greater than the filtered load of chloride. Almost all of the filtered load of both chemicals must be reabsorbed in the renal tubules in order to conserve total body electrolyte and fluid volume. We must consider this reabsorption in some detail. In the present context, only the movements across the tubular wall between tubular fluid and interstitial fluid need to be considered, not the cellular mechanisms which the tubular cells utilize to effect this transfer. The tubular wall is represented simply as a line separating the two fluids, as in Figure 3.5C. Sodium ions are the ions moved in greatest quantities across the tubular walls. The net tubular reabsorption of sodium is represented by a long arrow from tubular fluid to interstitial fluid. The accompanying chloride reabsorption is represented by a slightly shorter solid arrow in the same direction, the difference in length signalling that, since less chloride is filtered, there is necessarily less to be reabsorbed.

The total transfer of charge across the tubular wall must be zero; if this were not so, a large electrical potential would rapidly develop and prevent further net movement of charge. For every Mole of sodium ions transferred from tubular fluid to the renal interstitial fluid, there must be an accompanying Mole of charge to balance this. Whilst most of this is matched by chloride, the movement of other ions must make good the disparity. As shown by the solid arrows in Figure 3.5C, the contributors are bicarbonate, negatively charged, moving from tubular fluid to interstitial fluid, and potassium and hydrogen ions, both positively charged, being secreted into the tubular fluid and therefore moving in the opposite direction. Quantitatively, the sum of all the arrows, taking the polarity of the charge into account, is zero. (The transfer of potassium is actually more complicated than the diagram suggests; see section A.3.)

The balance shown in Figure 3.5C is disturbed when the extracellular hydrogen ion concentration changes. If this falls, the renal compensation involves a reduction in bicarbonate reabsorption and a reduction in hydrogen ion excretion (Figure 3.5C). These changes in composition of the urine result in the production of an alkaline urine as part of the physiological response to

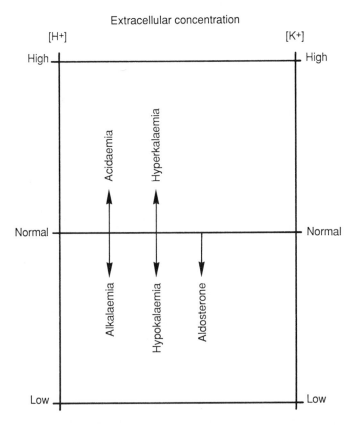

Figure 3.6. Diagram to illustrate the fact that changes in the extracellular concentration of hydrogen ions are usually in the same direction as those of potassium concentration and vice versa. Each horizontal line joins corresponding values of extracellular concentrations in a given subject at a given time. The arrows indicate the directions of change of this horizontal line in various conditions considered in the text.

rid the body of the alkaline load. To counterbalance these changes in tubular reabsorption and secretion, a transfer of positive charge from renal interstitial fluid to renal tubular fluid is mandatory. Potassium ions are available to carry this charge; they are therefore excreted in the urine in excess, at the expense of a lowering of the extracellular concentration of potassium ions. Since potassium ions secreted into the tubular fluid are lost to the body, there is depletion of the body's store of potassium.

There is a reciprocal behaviour in the renal tubule when a fall in potassium concentration in the extracellular fluid is the prime defect. The renal physiological mechanism is to retain potassium by reducing the loss of potassium in the urine. Reference to Figure 3.5C indicates that a reduction in the length of the arrow representing potassium reabsorption can be compensated, from an electroneutrality viewpoint, by an increase in hydrogen ion excretion and an increase in bicarbonate reabsorption. These movements render the renal interstitial fluid and hence the body as a whole alkaline; once again a lowering of extracellular potassium concentration leads to a lowering in its hydrogen ion concentration.

A summary of the inter-relationship between hydrogen ion concentration and potassium ion concentration in the extracellular fluid is shown in Figure 3.6. The horizontal line which moves up and down represents the fact that changes in the concentrations of the two species occur together in the same direction.

OVERVIEW OF HYDROGEN ION–POTASSIUM ION INTER-RELATIONSHIPS

The explanation for the changes in extracellular hydrogen and potassium ions occurring in parallel hinges on the principle of electroneutrality. This is a constraint which applies to all systems, animate and inanimate. Despite the fact that, in the context of acid–base physiology, the principle has been applied to the kidney, the whole argument depends on the physics of the system and is independent of the details of renal mechanisms. Whatever mechanism were operating for the maintenance of ionic composition of the extracellular fluid, it would necessarily incorporate the principle of electroneutrality.

ADVERSE EFFECTS OF CHANGES IN EXTRACELLULAR POTASSIUM CONCENTRATION

Hyperkalaemia. The most important effects are: high-peaked T wave in the ECG, heart block and cardiac arrest. **Hypokalaemia**. The most important effects are: failure of contractility of skeletal muscle, of smooth muscle (paralytic ileus) and of cardiac muscle.

SECTION 3.3 GASTRIC FUNCTION

The gastric juice contains acid at the highest concentration to be found anywhere in the body. It is therefore appropriate to consider the secretion of gastric juice and its influence on the acid–base status of the body.

Since the gastric juice of a normal person is strongly acid, secretion of acid by secretory cells in the stomach walls is against a very steep concentration gradient. Such secretion therefore requires a large amount of energy, derived from metabolism. The chloride concentration of gastric juice is also higher than that of the extracellular fluid and energy is also needed to pump this ion into the gastric juice. These active pumping processes produce a gastric juice which is slightly hyper-osmotic to the extracellular fluid, and water moves into the gastric juice passively as a result of the osmotic gradient (Friedman, 1975). This is the mechanism of secretion of the water content of gastric juice.

For every hydrogen ion secreted into the lumen of the stomach, a bicarbonate ion appears in the extracellular fluid of the stomach. This bicarbonate is drained by the gastric venous blood into the general circulation. During the period when a meal is being digested and much acid is being secreted into the stomach, the balancing transfer of bicarbonate into the body as a whole causes a transient physiological metabolic alkalosis. The flow of bicarbonate is called the **alkaline tide**. Later, when on balance more alkaline juice is being secreted into the intestinal tract in pancreatic juices, the balance is redressed, with an equivalent amount of acid being offloaded from the stomach into the body. This counteracts the alkalosis.

THE COMPOSITION OF GASTRIC JUICE

The osmolarity of gastric juice is similar to that of extracellular fluid, i.e. typically 300 milliosmols per litre. Since the electrolytes in gastric juice are monovalent, the osmolarity is made up from equal contributions from anions and cations (see Chapter 5). As shown in Figure 3.7, the anion composition is simple: chloride constitutes almost all of the 150 mM; the bicarbonate concentration at the low pH values pertaining in gastric juice is negligible. The concentrations of cations in gastric juice depend on its pH value. The left-hand graph of Figure 3.7 shows cation concentrations at various pH values from 0.9 units up to 2.0 units. At a pH of 0.9 units, the hydrogen ion concentration is 120 mM, leaving 30 mM to be made up by other cations. The potassium concentration of gastric juice is typically 10 mM, more than twice the value for extracellular fluid, so at this pH, the sodium concentration of gastric juice is 20 mM. At a pH value of 1.0 units, the $[H^+]$ is 100 mM and the $[Na^+]$ is 40 mM. If the pH of the gastric juice is less acid, at 2.0 units, then only 10 mM

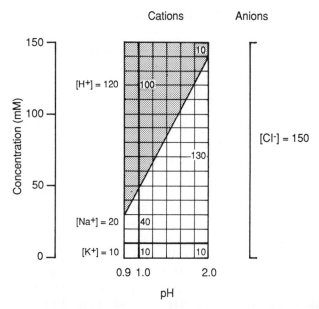

Figure 3.7. Bar diagram to show the composition of gastric juice. The anion composition is always of chloride at a concentration typically of 150 mM. The cation composition reflects the pH of the gastric juice, as the diagram illustrates. The potassium concentration is fairly constant at around 10 mM. The sodium and hydrogen ion concentration together provide the remaining 140 mM. The hydrogen ion concentration, shown as the shaded area, is 120 mM at pH = 0.9, 100 mM at pH = 1.0 and 10 mM at pH = 2.0. The respective concentrations of sodium ions are therefore 20, 40 and 130 mM.

is contributed by hydrogen ions and sodium ions comprise the majority of cations.

INTERACTIONS BETWEEN RESPONSES TO DISTURBANCES OF ACID–BASE PHYSIOLOGY AND OF FLUID VOLUME SUCH AS VOMITING

The effect of vomiting on acid–base status depends on whether there is a net loss of acid or of alkali in the vomitus. In cases where the movement of chyme from the stomach to the duodenum is obstructed, the resulting vomiting will be of gastric contents which are acid. In other cases of vomiting due, for instance, to irritation of the small bowel, the vomitus is a mixture of gastric contents, which are acid, and of duodenal contents, which are alkaline. In this situation the vomitus is acid if gastric juice predominates and alkaline if duodenal contents predominate.

VOMITING OF GASTRIC CONTENTS

This results in loss of hydrochloric acid from the body. Loss of acid is equivalent to gain of alkali, so the effect is a metabolic alkalosis. This is explained by reaction 1 in Table 3.7. Considering firstly the ions on the left, the sodium and chloride ions are quantitatively the major electrolytes in the extracellular fluid whilst the hydrogen ions and hydroxide ions come from the dissociation of water. The vomitus contains a high concentration of hydrochloric acid, which is thus lost from the body. This leaves behind sodium hydroxide. This shows that vomiting of hydrochloric acid solution is equivalent, from the point of view of acid–base physiology, to the addition of the strong alkali sodium hydroxide. Even with persistent vomiting, the high concentration of hydrogen ions in the gastric juice is maintained. In patients with pyloric obstruction, secretion of gastrin appears to be stimulated (Cohen and Kassirer, 1982, p. 257) so that the concentration of acid in the gastric juice and hence in the vomitus is maintained, typically at $100\,mM$ (pH = 1.0).

If the vomiting is moderate, the metabolic alkalosis calls up the physiological mechanisms which include renal excretion of bicarbonate. As a result, the urine is alkaline, an appropriate response since it rids the body of the alkali which is the origin of the acid–base problem. With persistent prolonged vomiting of gastric contents the loss of extracellular fluid becomes severe and may be life-threatening. Now, physiological mechanisms are called up to retain electrolytes in defence of the extracellular osmolarity and hence of extracellular fluid volume; the principal electrolytes concerned with this are sodium and chloride. There is conflict, as will emerge shortly, between the two sets of compensatory mechanisms. Retention of electrolytes and fluid takes precedence over acid–base homeostasis. This may lead to the condition of **paradoxical aciduria**, as the following account demonstrates.

Table 3.7

$$Na^+ + Cl^- + H^+ + OH^- \rightarrow H^+ + Cl^- + Na^+ + OH^- \qquad \text{reation 1}$$

| In extra-cellular fluid | From water | Lost in vomitus | Remains in extracellular fluid |

Abbreviated form

Extracellular electrolytes + Water

$$Na^+ + Cl^- + H^+ + OH^-$$

Lost in vomitus

leaving $Na^+ + OH^-$ in extracellular fluid.

In persistent vomiting of gastric contents, electrolytes are no longer available to the body from the alimentary tract to replace those lost in vomitus and in urine. The principal chemicals lost as a result of vomiting gastric contents are water, hydrochloric acid (as hydrogen ions and chloride ions) and sodium chloride (as sodium ions and chloride ions). The single electrolyte lost in greatest amounts is chloride, with the loss of hydrogen, sodium and potassium ions together balancing this. As a result of loss of chloride, the plasma chloride concentration falls from a normal value of 100 mM, and in severe cases to as low as 60 mM. The cumulative loss of sodium ions may be severe; even though the extracellular sodium concentration falls little, the fall in absolute amount is parallelled by a fall in extracellular fluid volume. The volume of plasma, a component of the extracellular fluid, is reduced with a consequent reduction in the volume of the circulating blood (hypovolaemia). This is potentially dangerous because, if sufficiently severe, it results in circulatory collapse and death.

With hypovolaemia, there is a reflex constriction of the afferent arterioles

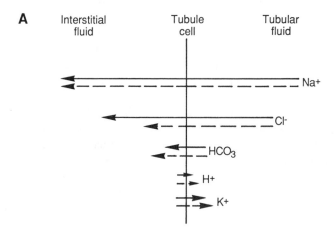

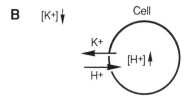

Figure 3.8. Movements of ions between the renal tubular fluid and the renal interstitial fluid: sodium, chloride and bicarbonate are reabsorbed, hydrogen and potassium ions are secreted. The bold arrows show movements in a normal kidney; the dashed arrows show the movements when the plasma chloride concentration is depressed.

in the renal corpuscles in the kidney with a consequent reduction in glomerular filtration rate. Less water and electrolytes are filtered and less reabsorption is needed than normal to retain electrolytes. Additionally, to compensate for the hypovolaemia, powerful compensatory mechanisms for water and electrolyte retention are active; these mechanisms include the renal retention of sodium ions.

The low concentration of chloride in the plasma results in a relatively small filtered load of chloride ions by comparison with sodium ions. In the renal tubular fluid, there is less chloride than usual available for reabsorption to balance the reabsorption of sodium. This is shown in Figure 3.8A, the disparity between the lengths of the dashed sodium and chloride arrows being greater than normal. The body cannot respond by reducing sodium reabsorption, since sodium reabsorption is required in the defence of the extracellular fluid volume. The only mechanism available to the kidney is to increase the movement of HCO_3^- from tubular fluid to interstitial fluid, and the secretion of H^+ and K^+ into the tubular fluid; these transfers are indicated by the dashed arrows in Figure 3.8A. From the point of view of acid–base physiology, the renal retention of HCO_3^- exacerbates the alkalosis which already exists. There is increased renal excretion of H^+ yielding an acid urine. Excretion of acid in the urine results in loss of acid from the body, leaving the extracellular fluid yet more alkaline. Again, the metabolic alkalosis is exacerbated. The plasma bicarbonate frequently reaches values of 60 mM; values as high as this occur almost exclusively in patients with gastric alkalosis (Cohen and Kassirer, 1982, p. 257).

From the point of view of potassium balance, there is increased renal excretion of potassium, loss of potassium in the vomitus and no potassium being delivered for absorption in the alimentary tract. All these factors contribute to a severe depletion of the body's total potassium content. Yet another factor contributes to potassium loss. A drop in volume of the circulating blood leads to aldosterone secretion via the renin-angiotensin mechanism which, in turn, promotes sodium reabsorption in the renal tubule; this contributes further to excessive renal loss of potassium and hydrogen ions. The acidity of the urine is inappropriate as a response to metabolic alkalosis, but the preservation of electrolyte and fluid volume takes precedence over the acid–base disturbance. These various effects all combine to yield a positive feedback system driving the metabolic alkalosis which, if not treated, reaches lethal levels in a few days.

As already explained, the renal response to loss of extracellular fluid volume results in a low plasma potassium concentration. As shown in Figure 3.8B, the lowering of the extracellular concentration of potassium results in an increased potassium concentration gradient from intra- to extracellular fluid. Consequently, potassium diffuses from the intracellular to the extracellular compartment. To maintain electrochemical neutrality, hydrogen ions move in the opposite direction, from extracellular to intracellular fluid. The extracellular metabolic alkalosis is even further exacerbated whilst the intracellular fluid becomes more acid than usual. There is an intracellular acidosis accompanying the extracellular alkalosis.

The potassium leaving cells and entering the extracellular fluid partially replaces the potassium lost into the renal tubular fluid and is of itself just a redistribution of potassium within different fluid compartments in the body. It is not lost from the body. However, the potassium moving from the intra- to the extracellular compartment bolsters the extracellular concentration of potassium, and, by the electroneutrality effect described earlier, becomes available for excretion in the urine, thereby being lost to the body entirely. The overall effect is a dramatic lowering of total potassium content of the body.

OUTLINE OF TREATMENT

Cautious intravenous infusion of isotonic sodium chloride solution improves the patient's condition. Although sodium chloride solution is neutral, the provision of the principal extracellular electrolytes provides the necessary osmotically active material to return the extracellular fluid volume towards normal. This switches the drive from mechanisms for retention of electrolyte to those for correction of metabolic alkalosis. The body puts its own house in order so far as the acid–base problems are concerned and the situation is saved.

APPENDIX 3

A.1 Renal compensation for respiratory acidosis

Renal compensation returns the plasma hydrogen ion concentration towards normal. In some cases the compensation is complete and in a few over-compensation occurs. This indicates that the physiological mechanism is not a simple negative feedback loop with the error signal being the increase in plasma hydrogen ion concentration; so long as such a mechanism does not oscillate, it never completely restores normality because, were it to do so, the error signal would be eradicated. Still less can such a feedback mechanism lead to over-compensation.

The details of the physiological mechanism for compensation are not understood. Possible mechanisms include: (a) renal reabsorption of bicarbonate as a result of the increased P_{CO_2} and (b) resetting of the set point of a mechanism dependent on the hydrogen ion concentration.

A.2 Causes of metabolic acidosis

The causes can be grouped into four groups:

1. increase in hydrogen ion load taken by mouth (e.g. ingestion of ammonium chloride) or by injection;
2. excessive loss of bicarbonate ions from the body through the gastrointestinal tract (e.g. diarrhoea, pancreatic or biliary drainage);
3. increase in hydrogen ion load from endogenous sources;
4. decrease in the kidney's ability to excrete acid, due to carbonic anhydrase inhibitors or renal disease.

A.3 Renal handling of potassium

At the glomerulus, potassium ions are freely filtered with some 900 mMoles being filtered in 24 hours. An amount equal to the filtered load is reabsorbed. About 100 mMoles of potassium is also secreted into the tubular fluid and in normal circumstances this amount is excreted in the urine. The net movement across the tubular wall is thus around 800 mMoles in 24 hours from tubular fluid to renal interstitial fluid. In each of Figures 3.5C and 3.8, the arrow indicating transfer of potassium relates to secretion. The total transfer of potassium across the tubular wall is in the direction opposite to that shown for secretion. If this total transfer had been used, the analysis would have been more rigorous but the argument would have become unnecessarily complicated. The interested reader is encouraged to work through the more rigorous analysis as an exercise.

Assessment of acid–base status | 4

SECTION 4.1 STANDARD BICARBONATE, BASE EXCESS AND BUFFER BASE

MEASUREMENT OF ACID–BASE STATUS

In the previous chapter, we considered respiratory and metabolic disorders of acid–base physiology. In respiratory disorders, the immediate change is in the P_{CO_2}, with a move away from the normal blood composition along the normal blood line (Figure 3.1, arrow N A). In metabolic disorders, the immediate change is a move to a new blood line (Figure 3.4, arrow N A). When a disorder of respiration is the primary factor, compensation involves secondary changes in renal handling of bicarbonate (e.g. retention of bicarbonate in hypoventilation and a resultant move to a new blood line, Figure 3.1, arrow A B). When a disorder of metabolism is the primary factor, compensation involves secondary changes both in ventilation (e.g. Figure 3.4, arrow A B) and in renal handling of bicarbonate (e.g. Figure 3.4, arrow B C). Compensation introduces features in addition to the primary moves.

In general, if the primary disturbance causes an excess of unwanted acid in the body, the compensation involves addition to the body of extra alkali to neutralize the acid. A primary acidosis calls up in response a physiological alkalosis. The disorder of acid–base physiology is then a mixture of these two components which operate in opposite directions. A primary acidosis is accompanied by a secondary (compensatory) alkalosis and vice versa. It is clearly important to unravel the different components of an acid–base disorder from an examination of a sample of arterial blood.

In any disorder of acid–base physiology, there are liable to be two components, respiratory and metabolic, and the contribution of each must be identified. For the respiratory component, the arterial P_{CO_2} immediately indicates any deviation from normality (Table 4.1A). If this is within normal limits, there is no respiratory component to the acid–base disorder.

The metabolic component is indicated by the shift in the blood line. Consequently, it is important to know whether the measurements made on an arterial

Table 4.1

A. *Components of an acid–base disorder*

   Respiratory: indicated by arterial P_{CO_2}.

   Metabolic: indicated by a shift in the blood line.

B. *Standard bicarbonate*: the concentration in mM of bicarbonate in the plasma of oxygenated whole blood equilibrated with a P_{CO_2} of 40 mmHg at 37°C.

   Measurement:
   1. At 37°C, equilibrate the sample of arterial blood with a gas mixture containing CO_2 at a partial pressure of 40 mmHg.
   2. Measure $[HCO_3^-]$.

   Values: typical normal value: 24 mM.
   Above 26 mM indicates metabolic alkalosis
   Below 22 mM indicates metabolic acidosis.

---

blood sample indicate that the patient is on the normal blood line (as, for example, in uncompensated respiratory acidosis, point A in Figure 3.1) or has moved away from this line (as in compensation for primary respiratory disorder, point B in Figure 3.1). Consider point A in Figure 3.1. Points A and N differ in $[CO_2]$, $[HCO_3^-]$ and $[H^+]$ so none of these alone indicates that point A lies on the normal blood line. A combination of the three values must be used. There are three commonly used combinations, each with its advantages and disadvantages. These are the 'standard bicarbonate', base excess and buffer base or total buffer base which are considered in turn, starting with the standard bicarbonate.

INDICATORS OF THE METABOLIC COMPONENT

The standard bicarbonate

The respiratory component of an acid–base disorder is due directly to altered P_{CO_2}. So its effects can be reversed by equilibrating the sample of blood at 37°C with a gas containing carbon dioxide at a partial pressure of 40 mmHg, the normal value. Then the bicarbonate concentration is measured. This is called the 'standard bicarbonate' because it is measured at a standard partial pressure of carbon dioxide. The standard bicarbonate is thus defined as the bicarbonate concentration in millimoles per litre of plasma of oxygenated whole blood which has been equilibrated at 37°C with a gas mixture having a P_{CO_2} of 40 mmHg. A typical value for the standard bicarbonate in a normal subject is 24 mM.

The procedure is shown in graphic form in Figure 4.1. Suppose that a blood

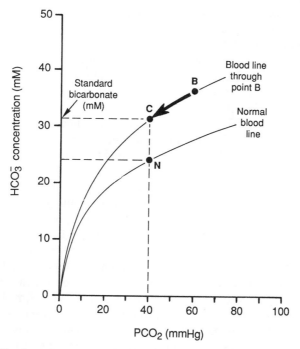

Figure 4.1. Bicarbonate concentration as a function of P_{CO_2}. The construction in this graph shows the rationale behind the measurement of the 'standard bicarbonate'.

sample has a composition placing it at point **B**. Since the P_{CO_2} is above 40 mmHg, there is clearly a respiratory component, the magnitude of which is indicated by the size of the increase in P_{CO_2}. Since point **B** lies above the normal blood line, which is included for reference, there is also a metabolic component; we wish to assess the magnitude of this metabolic component. Equilibration with a gas mixture containing carbon dioxide at a partial pressure of 40 mmHg carries the blood along its own blood line to point **C**. This is the point on the subject's blood line corresponding to the partial pressure of carbon dioxide pertaining when the contribution of the respiratory disturbance has been artificially removed. There is a significant deviation of the standard bicarbonate from the normal value of 24 mM indicating a metabolic component to his acid–base disturbance. Because the bicarbonate concentration is high at standard P_{CO_2}, there is a metabolic alkalosis. The standard bicarbonate gives a measurement of the magnitude of this metabolic component.

The range of values of the standard bicarbonate measured from blood samples in people in a normal healthy population is between 22 and 26 mM; a value outside this range indicates a metabolic disturbance of acid–base status. A standard bicarbonate above 26 mM indicates metabolic alkalosis as a component for the acid–base disturbance; a standard bicarbonate below 22 mM indicates metabolic acidosis.

The need for a better measure of the metabolic component of an acid–base disorder. At first sight, we might think that the deviation of the standard bicarbonate from the value for normal blood is all the information that we need about the non-respiratory component of an acid–base disorder. However, the change in standard bicarbonate underestimates the non-respiratory component of the acid–base disorder. To illustrate this, consider uncompensated metabolic alkalosis, produced by the addition of alkali, such as sodium hydroxide, to the blood. Each of the buffer acids in the blood (protein buffer acid and CO_2) buffers some of the added alkali as shown in the two chemical reactions in Table 4.2A. Protein buffer acid combines with some of the alkali to yield water and protein buffer base; CO_2 from metabolism combines with most of the rest of the alkali to yield bicarbonate. The result is an increase in concentration both of non-bicarbonate buffer base Pr^- and of bicarbonate.

The change in acid–base status in uncompensated metabolic alkalosis is represented by the move from point N to point C in Figure 4.1. The increase in bicarbonate concentration is evident from the upward movement. This point C has a P_{CO_2} coordinate of 40 mmHg, so the value of the bicarbonate at this point is also the standard bicarbonate for this sample of blood. The rise in standard bicarbonate only estimates the contribution of the CO_2-bicarbonate system (reaction 2 in Table 4.2A), ignoring that of the non-bicarbonate buffer (reaction 1 in Table 4.2A). The relative amount of buffering provided by the two systems varies in different conditions. To measure the excess alkali in blood at point C, it is necessary to measure the increase of both $[Pr^-]$ and $[HCO_3^-]$ in the blood. This is called the **base excess**. To measure the base excess directly, 'back titration' must be used.

Base excess

The 'base excess' is the change from normal of the sum of the concentration of bicarbonate and non-bicarbonate buffer base ($[HCO_3^-] + [Pr^-]$). The steps involved in the measurement are shown in Table 4.2B. The first step is to take a measured volume of the patient's blood and to equilibrate it at 37°C with a gas mixture containing carbon dioxide at a partial pressure of 40 mmHg. This removes any respiratory component of the acid base disorder.

The problem to be solved is to estimate the amount of non-respiratory alkali which has been added to the blood. The hydrogen ion concentration itself would only give us the answer if there were no buffer. The presence of blood buffers reduces the change in $[H^+]$ by providing most of the hydrogen ions for the strong alkali. The pH of blood from a normal subject is 7.4. So in order to estimate the excess of alkali in an arterial blood sample from a patient with metabolic alkalosis, enough strong acid, such as hydrochloric acid, is added to bring the pH back exactly to 7.4. By this stage, the strong

Table 4.2

A. *Buffering of added alkali*

$$OH^- \;+\; HPr \;\to\; H_2O + Pr^- \qquad \text{reaction 1}$$
$$OH^- \;+\; CO_2 \;\to\; HCO_3^- \qquad \text{reaction 2}$$

Added	Blood
alkali	buffer
	acids

B. *Base excess*: the change from normal of $([HCO_3^-] + [Pr^-])$.

Measurement:
1. Equilibrate the sample of arterial blood with gas containing CO_2 at $P_{CO_2} = 40\,mmHg$.
2. Titrate with strong acid or alkali to a pH of 7.4.

Base excess is measured in mM.
Normal range is $\pm 2.5\,mM$.

---

acid has exactly counteracted the metabolic alkali. This is the principle of 'back-titration', since the blood has been titrated back to the pH of 7.4. Quantitatively, the base excess is the amount of strong acid, in mM, which has to be added to one litre of blood to bring its pH back to 7.4. Base excess is thus measured in units of mM. Nomogram methods are also available, as described later in this chapter. The base excess is defined operationally as the amount of strong acid required to titrate the pH of the blood to 7.4 at a P_{CO_2} of 40 mmHg and at 37°C.

An excess of base in the blood is referred to, naturally enough, as a **base excess**. When there is an excess of non-respiratory acid in the blood, alkali is added to bring the pH to 7.4 units. There is then said to be a **base deficit**. Base deficit is sometimes called **negative base excess**. This apparently confusing nomenclature simplifies reports from the blood laboratory in hospitals. In laboratory forms reporting on the acid–base status of a patient, this information is usually expressed as base excess, which may be positive or negative. The meaning of base excess is indicated graphically in Figure 4.2. With excess alkali in the blood there is a base excess whereas with excess non-respiratory acid there is base deficit, or 'negative base excess'.

The base excess is an estimate of the magnitude of the non-respiratory component of an acid–base disturbance. In a population of normal individuals with no disturbance of acid–base physiology, the base excess is normally in the range of $+/-2.5\,mM$.

Buffer base or total buffer base

The summed concentration of bicarbonate and non-bicarbonate buffer base $([HCO_3^-] + [Pr^-])$ is called the **total buffer base** or, more usually, the **buffer base** (Table 4.3A). For blood of normal composition in equilibrium with gas

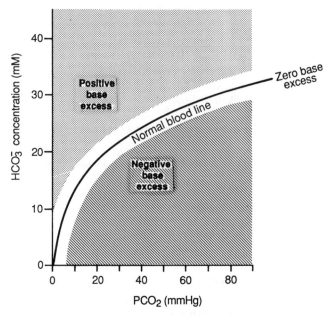

Figure 4.2. Graph to show the regions of the acid–base status plot corresponding to positive and to negative base excess. The unshaded area around the 'normal blood line' gives an approximate indication of the variations in base excess to be found in a normal healthy population.

Table 4.3

A. *Buffer Base (or Total Buffer Base)* is $([HCO_3^-] + [Pr^-])$
   Typical value for normal blood is 48 mM.

B. *Effect of an increase in* P_{CO_2}

   Added
   $$CO_2 + H_2O \rightarrow HCO_3^- + H^+ \qquad\qquad \text{reaction 1}$$
   $$H^+ + Pr^- \rightarrow HPr \qquad\qquad\qquad \text{reaction 2}$$
   So $([HCO_3^-] + [Pr^-])$ is unchanged.

containing carbon dioxide at 40 mmHg, the concentrations of bicarbonate and of non-bicarbonate buffer base are approximately equal. The value of total buffer base in normal blood is thus around 48 mM. As with the base excess, the buffer base is a measure of the metabolic component of a disturbance of acid–base physiology; it is uninfluenced by any respiratory component as will be shortly demonstrated. The buffer base is measured by chemical analysis of the blood (Singer and Hastings, 1948). It is one of the parameters which is read off from the Siggaard-Andersen nomogram, to be described later in this chapter. Section A.1 briefly describes the chemical estimation of total buffer base.

THE EFFECT OF CHANGES IN P_{CO_2} CONCENTRATION ON BLOOD CHEMISTRY

If the P_{CO_2} with which a sample of blood is equilibrated increases, the reactions which take place are as shown in Table 4.3B. One molecule of CO_2 reacting with water yields one bicarbonate ion and one hydrogen ion (Table 4.3B reaction 1). Almost all of the hydrogen ions so released are buffered by non-bicarbonate buffer base Pr^- to yield protein buffer acid HPr (Table 4.3B reaction 2). To a close approximation, the hydration of one molecule of carbon dioxide causes the release of one ion of HCO_3^- and the withdrawal of one ion of Pr^-, the sum of ($[HCO_3^-] + [PrA^-]$) remaining constant. Non-bicarbonate buffer base flips over to bicarbonate buffer base without changing the sum of the two, i.e. the total buffer base ($[Pr^-] + [HCO_3^-]$) is unaltered. This example demonstrates the general rule that, whilst the relative amounts of each of the two components, bicarbonate and non-bicarbonate, of the total buffer base depend on the partial pressure of carbon dioxide, the concentration of total buffer base is independent of the P_{CO_2}.

This characteristic behaviour of the blood in response to changes in the partial pressure of carbon dioxide is incorporated in the plot of acid–base status in Figure 4.3A. The normal blood line is shown in the usual fashion. For a given P_{CO_2}, the bicarbonate concentration is indicated by a vertical arrow between the x-axis and the blood line. Above the blood line, a second arrow continues vertically upward to indicate the non-bicarbonate buffer base $[Pr^-]$. At a physiological P_{CO_2} and pH, the $[HCO_3^-]$ and $[Pr^-]$ are approximately equal at 24 mM each. The total height of these two vertical arrows for a given P_{CO_2} is the total buffer base. This is shown on the scale to the right of the graph. If the tops of all the arrows representing total buffer base at various values of P_{CO_2} are joined, this gives a straight line with zero slope, i.e. a value of total buffer base which is independent of the P_{CO_2}. It is exactly this behaviour of total buffer base which makes it so useful as an indicator of acid–base status. It indicates only the metabolic component of an acid–base disorder, being insensitive to any change in the respiratory component.

WHY ARE BASE EXCESS AND BUFFER BASE BOTH USED?

The concentration of total buffer base depends on the overall buffering power of the blood. Since much of the buffering is due to haemoglobin, in an anaemic person the total buffer base is low. An anaemic person with no disturbance of acid–base physiology will normally operate at a normal P_{CO_2} and normal

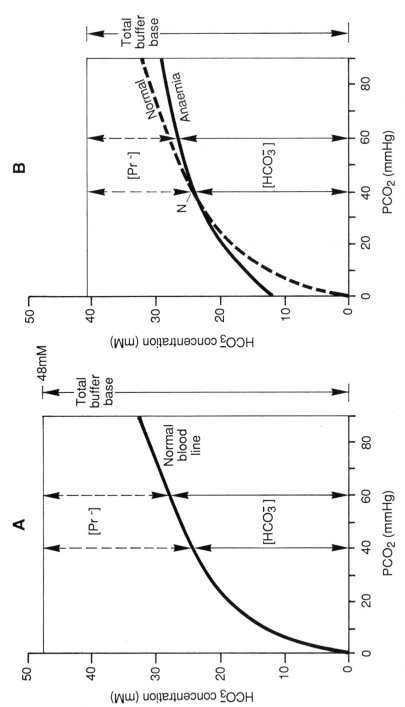

Figure 4.3. A. Normal blood: the 'normal blood line' is plotted with axes as indicated below and to the left. The full arrows show the value of the [HCO₃⁻] at the relevant Pco₂. Above each full arrow is a dashed arrow indicating the concentration of non-bicarbonate buffer base. The two arrows together indicate the magnitude of the total buffer base. B. A similar plot for anaemic blood.

blood $[H^+]$ so, by the Henderson-Hasselbalch relation, the plasma bicarbonate will have a normal value. The blood line for the anaemic subject thus passes through the same 'typical arterial point' N on the acid base plot as that for a non-anaemic subject (Figure 4.3B). The standard bicarbonate is normal. Since at a partial pressure of 40 mmHg the blood pH of the anaemic subject is 7.4, his base excess is zero. The only measurement which we have considered in the context of acid–base physiology and which brings out the deficient buffering power of the blood is the total buffer base, hence its value.

The basic deficiency is the lower-than-normal concentration of non-bicarbonate buffer base. When the P_{CO_2} with which the blood is equilibrated is changed, the inferior buffering power of anaemic blood is expressed. The slope of the blood line is less than for non-anaemic blood (Figure 4.3B). This is matched by lower values for non-bicarbonate buffer base at all P_{CO_2} values. Correspondingly, the total buffer base is less than 48 mM, because of a relative lack of non-bicarbonate buffer base.

This point is taken up again later in the chapter because the haemoglobin concentration appears on the Siggaard-Andersen nomogram.

SPECIFICATION OF ACID–BASE STATUS

Figure 4.4 shows how the P_{CO_2}-bicarbonate plane may be divided into four areas. The boundaries are the normal blood line and a vertical line with P_{CO_2} coordinate of 40 mmHg. Each area has a different combination of the change in P_{CO_2} and of the base excess. This allows us to specify, for any point, the P_{CO_2} quantitatively and the base excess qualitatively. For a quantitative specification of base excess, all that can be stated at this stage is that the further away the point representing arterial plasma is from the normal blood line, the greater the base excess or base deficit. It will be shown later that a refinement of graphical representation allows the specification of the base excess quantitatively.

In previous chapters, respiratory and non-respiratory disorders of acid–base physiology have been described. In this chapter, methods of measuring the respiratory and non-respiratory components of a disturbance of acid–base physiology have been considered. We now consider these two groups of ideas together.

In the initial stages of a disturbance of acid–base physiology, the condition is uncompensated, which essentially means that chemical buffering alone is operating. At this stage therefore, there is only one component to the disorder. This component is respiratory in respiratory disorders and metabolic in metabolic disorders. In the respiratory disorder the subject moves along the normal blood line. The partial pressure of carbon dioxide is abnormal but the base excess, the measure of metabolic component, is zero. This is shown in Table 4.4A. For an uncompensated metabolic disorder (Table 4.4C), it is the

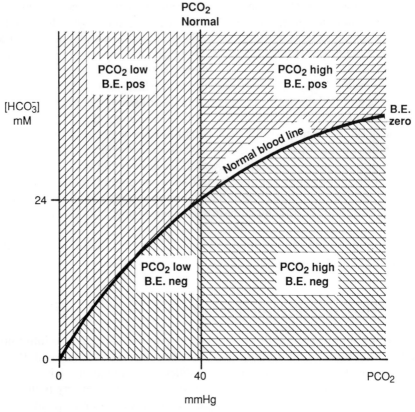

Figure 4.4. Bicarbonate concentration as a function of P_{CO_2}. The plane is divided into four areas with the different combinations of P_{CO_2} and base excess, as indicated.

Table 4.4

		Respiratory component P_{CO_2}	Metabolic component Base excess
A. Respiratory acidosis	Uncompensated	High	Zero
B.	Compensated	High	High
C. Metabolic alkalosis	Uncompensated	Zero change	High
D.	Compensated	High	High

base excess which is abnormal and the partial pressure of carbon dioxide which is unaltered. In the immediate response to acid or alkaline load, be it of respiratory or metabolic origin, the blood chemistry is unambiguous.

Compensation renders the situation more complex. In respiratory acidosis, for instance, compensation involves the retention of bicarbonate by the kidneys as illustrated in Figure 3.1. This results in the subject moving to a new blood line indicating the addition of a metabolic component to an initially purely respiratory effect. The compensation means that there is now a positive base excess in addition to the raised partial pressure of carbon dioxide. So in compensated respiratory acidosis, there are both respiratory and metabolic components. The respiratory component is the primary disorder and the metabolic component is the secondary effect. The primary respiratory component is in the direction of acidaemia and the secondary metabolic response component is in the direction of restoration of pH, i.e. in the direction of alkalaemia. In compensation, therefore, as shown in Table 4.4B, both the partial pressure of carbon dioxide and the base excess are raised.

Let us now consider what happens when the effect which was secondary in the primary respiratory disorder is instead the primary disorder. This would then be a primary metabolic alkalosis, as in the vomiting of gastric contents. In the uncompensated condition, there is a positive base excess with a normal partial pressure of carbon dioxide already noted (Table 4.4C). The respiratory compensation is hypoventilation, brought about by the partial withdrawal of the normal stimulus of hydrogen ions to the peripheral chemoreceptors. The partial pressure of carbon dioxide rises, adding a respiratory component to the acid–base disorder (Table 4.4D).

In both cases B and D, the subject has a high partial pressure of carbon dioxide and a positive base excess. From the chemistry of blood taken at this stage alone, it is impossible to diagnose the path along which the condition has been reached. Information about the previous pathway must be sought both from the account given by patient and relatives of how the problems first started, from consideration of clinical signs and from earlier analyses of the blood biochemistry when these are available. The laboratory result should never be considered in isolation. It should be used as an indicator of the progress of a dynamic process in a patient.

Further generalizations can be made in relation to the division of the acid–base status chart into four areas. Points on the normal blood line correspond to uncompensated respiratory disorders; points on the vertical line $P_{CO_2} = 40\,\text{mmHg}$ correspond to uncompensated metabolic disorders. Transfer of information from Table 4.5 onto Figure 4.5 indicates that the upper right region delineates the area corresponding either to compensated respiratory acidosis or to compensated metabolic alkalosis. Similarly, the lower left region indicates either compensated respiratory alkalosis or compensated metabolic acidosis. This leaves the upper left and lower right quadrants unoccupied. Points falling within these regions are indicative of dual primary disorders of

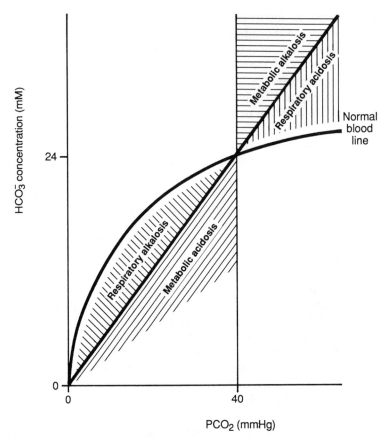

Figure 4.5. Similar to Figure 4.4, with indications of the regions of the plane corresponding to particular disorders of acid–base physiology.

acid–base physiology. The upper left quadrant indicates primary respiratory alkalosis and primary metabolic alkalosis occurring together. The lower right quadrant indicates primary respiratory acidosis and primary metabolic acidosis.

SECTION 4.2 THE SIGGAARD-ANDERSEN NOMOGRAM FOR THE PARAMETERS OF ACID–BASE STATUS

This section considers the Siggaard-Andersen nomogram for reading off the parameters of acid–base status such as the base excess. It has previously been noted that, although the P_{CO_2} and $[HCO_3{}^-]$ values for a sample of arterial blood define a point on the chart of acid–base status, they do not by themselves provide enough information to identify the magnitude of the base excess, total buffer base or standard bicarbonate. For each of these measures of the metabolic component of an acid–base disturbance, a chemical procedure such as back-titration is performed which allows for the buffering properties of the blood.

If it were known that the blood sample had the same buffering properties as some standard blood which had been exhaustively studied in the laboratory beforehand, it would be possible to attach to each point on the acid–base chart its base excess, total buffer base and standard bicarbonate. With such a labelling, it would suffice to identify the point on the chart for a sample of blood from a patient and all these parameters would be provided without the need for time-consuming back-titrations etc. This would be a considerable advantage in providing the best service for a patient with a severe disorder of acid–base balance and needing emergency treatment. It will soon emerge that two points must be determined to allow for the fact that different samples of blood have quantitatively different buffering capacities.

This approach has been successfully exploited, notably by Siggaard-Andersen (1964). He investigated the buffering behaviour of samples of blood from many normal subjects and many patients with disturbances of acid–base balance and found sufficient consistency of the blood lines from different subjects that he could forecast the behaviour of blood from a new subject. The procedure is as follows. Two measurements are made on the sample of arterial blood. Firstly, the pH and P_{CO_2} are measured on the sample as it is taken, to indicate its state *in vivo*. Secondly, the blood sample is equilibrated with a gas mixture containing CO_2 at a different partial pressure from that in the blood when taken, for example a partial pressure of 40 mmHg may used if the *in vivo* P_{CO_2} is high. The pH is measured. This gives two points on the blood line for this sample of blood and provides enough information to specify precisely the acid–base status of the blood, as will now be shown.

In the representation of acid–base status used so far in this book, $[HCO_3^-]$ has been plotted as a function of P_{CO_2}. This choice has advantages but, in the context of constructing a nomogram, it has the disadvantage that the blood line is curved, not straight. It often happens that a curved relationship can be transformed approximately into a straight line by using a logarithmic plot of the variables. Such proves to be the case for acid–base variables. Siggaard-Andersen chose to plot log $[P_{CO_2}]$ as a function of pH (which is $-\log[H^+]$).

In order to understand this plot, the first step is to identify contours representing constant bicarbonate concentration; it turns out (section A.2) that, for a given concentration of bicarbonate, such a contour has a slope of -1. Different concentrations of bicarbonate yield parallel lines; Figure 4.6 shows the three dashed straight lines corresponding to bicarbonate concentrations of 12 mM, 24 mM and 48 mM. In Chapter 2, it was shown that a line of constant bicarbonate concentration corresponds closely to the behaviour of a solution with no buffering power. Iso-pH lines are vertical lines on this plot. A solution which is extremely well buffered at pH 7.4 would yield a line close to the one

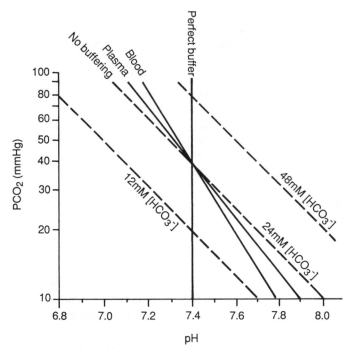

Figure 4.6. Pco_2 (logarithmic scale) as a function of pH. Points corresponding to a constant value of bicarbonate concentration lie on a line with a slope of -1; three such lines are shown as the interrupted lines in the figure. These lines also correspond to the relationships yielded by solutions with zero buffering power; the line representing a bicarbonate concentration of 24 mM is labelled 'no buffering'. As shown in this diagram, blood and plasma yield relationships which are close to straight lines over the whole physiological range of values of pH and Pco_2.

labelled 'perfect buffer'. Any real buffer solution such as blood or separated plasma will exhibit an intermediate relationship. A good buffer will give a buffer line which is almost vertical whereas a poor buffer will give a buffer line whose slope is closer to -1. Comparing blood and separated plasma, the separated plasma, being the poorer buffer of the two, has a slope nearer to -1 than the line for blood. If the plasma is derived from normal blood and is separated from the erythrocytes at a Pco_2 of 40 mmHg, the lines for blood and plasma cross at the physiological point (pH = 7.4, Pco_2 = 40 mmHg), as shown in Figure 4.6.

A strong acid, such as hydrochloric acid, is now added to normal blood and to normal separated plasma. In each case, 20 mM acid is added to one litre of blood or plasma. Each fluid now has a base excess of -20 mM. Next both fluids are equilibrated with a gas mixture containing carbon dioxide at a partial pressure of 40 mmHg and the pH values are measured. Although the blood is a good buffer, its pH falls; its status is shown by point A in Figure 4.7.

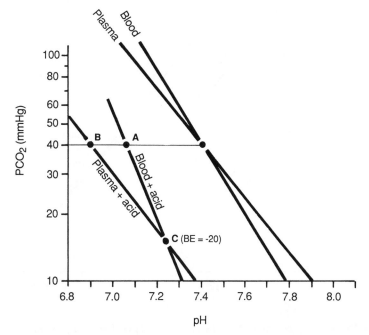

Figure 4.7. P_{CO_2} (logarithmic scale) as a function of pH. Comparison of the behaviour of normal blood and of plasma separated from normal blood at $P_{CO_2} = 40$ mmHg with blood or plasma to which fixed acid has been added, 20 mM per litre of blood or plasma.

The separated plasma is an inferior buffer so that the pH falls further than for blood; the status of the acidotic plasma is shown by point B in Figure 4.7. If the P_{CO_2} is varied and the resultant pH is measured, each fluid gives a relationship similar to that of its partner obtained before adding the strong acid but now the two lines cross at point C, with a value of P_{CO_2} below 40 mmHg. This point is labelled -20, to indicate the base excess of this blood and plasma.

In Figure 4.8, the data obtained so far has been re-plotted, together with measurements made when a similar procedure is performed as in the last paragraph, except that base has been added instead of acid. This gives the three pairs of intersecting straight lines. The points of intersection are labelled with the base excess values. When the procedure is repeated for intermediate values of added acid or base, the intercepts of the lines for blood and separated plasma give the curved dashed scale labelled in units of base excess.

By this stage, it has become apparent that this construction can be used as a nomogram to read off directly the base excess of a sample of blood, even if it is anaemic. For a sample of arterial blood, two points are plotted corresponding to the two values of P_{CO_2} with which the blood sample is equilibrated, as described above. These two points are joined with a straight line and the inter-

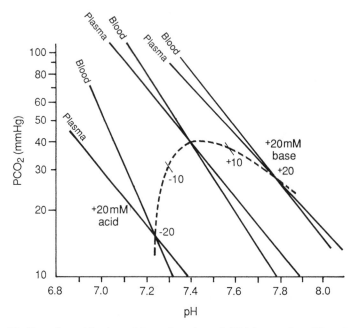

Figure 4.8. P_{CO_2} (logarithmic scale) as a function of pH. Three pairs of lines for blood and plasma, each pair corresponding to a particular value of base excess. The points of intersection of the pairs of lines are shown as the dashed curve, calibrated in units of base excess.

section of this blood line with the base excess scale gives directly the value of the base excess of the blood sample.

The foregoing has prepared the reader for the full Siggaard-Andersen nomogram shown as Figure 4.9. The scales for 'buffer base' and standard bicarbonate are constructed with reasoning similar to that for the 'base excess' scale. A pair of lines corresponding to blood and plasma, each with the same base excess, gives different values for the (total) buffer base. This is because total buffer base is a reflection of the buffering capacity, which is greater for blood than for separated plasma.

THE EFFECT OF HAEMOGLOBIN CONCENTRATION

When blood is taken from a patient for examination of acid–base status, the haemoglobin concentration of the blood is also measured routinely. Blood from an anaemic patient, being less well buffered than normal blood, yields a line on the Siggaard-Andersen plot with a slope nearer to -1 than normal blood. To calculate the contribution of the anaemia, it is necessary to introduce the concept of 'normal buffer base'. The **normal buffer base** of a sample of

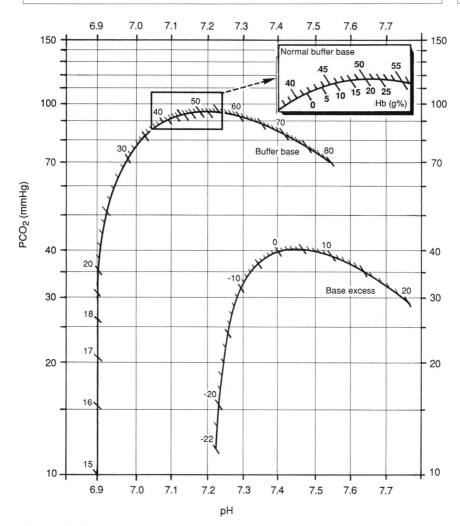

Figure 4.9. P_{CO_2} (logarithmic scale) as a function of pH. A re-drawing of the Siggaard-Andersen nomogram, derived, with permission, from the original nomogram published in 'The acid–base status of the blood'. Volume 15, Supplement 70, *Scandinavian Journal of Clinical and Laboratory Investigation*, 1963. The use of the nomogram is described in the text. The inset, upper right, is an enlargement of the part of the 'buffer base' curve used to derive an estimate, based on the buffering capacity of the blood, of the haemoglobin concentration of the patient's blood.

blood is the total buffer base of the blood when any changes contributed by a disturbance of acid–base balance have been removed. The normal buffer base for normal blood with a haemoglobin concentration of 15 g% is 48 mM.

To estimate the normal buffer base for a sample of blood giving a particular blood line, the procedure is to read off the corresponding base excess and the buffer base values. When there is a metabolic disorder of acid–base balance, the total buffer base comprises the normal buffer base plus or minus the contribution provided by the acid–base disturbance; this latter component is indicated by the base excess. As a formula, this gives:

Observed buffer base = normal buffer base + observed base excess.

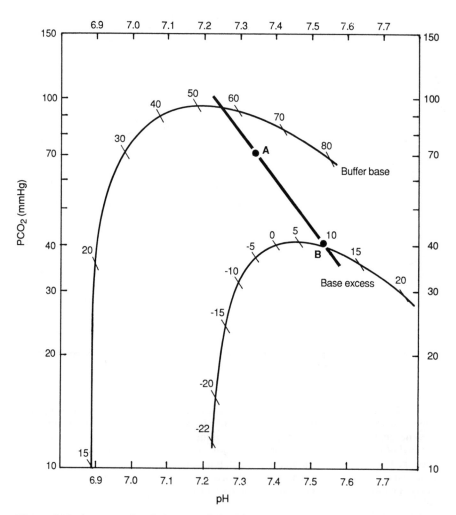

Figure 4.10. An example of the use of the Siggaard-Andersen nomogram. Arterial blood from a patient had a P_{CO_2} of 71 mmHg and a pH of 7.34; this is plotted as point A. The blood sample was then equilibrated with a gas mixture containing CO_2 at a partial pressure of 40 mmHg; the pH was then 7.54, giving point B. These points are joined by the straight line shown in bold. This allows us to read off the base excess and the buffer base values.

or

Normal buffer base = observed buffer base − observed base excess.

This last formula allows one to calculate the normal buffer base from values read from the Siggaard-Andersen nomogram. This normal buffer base value is then identified on the part of the buffer base curve against which is plotted the haemoglobin subscale and the haemoglobin is read off.

Suppose, for example, the blood line of a patient gave these values:

Buffer base	26 mM
Base excess	− 20 mM

Then Normal buffer base = 26 + 20 = 46 mM.

This value on the buffer base curve of the Siggaard-Andersen plot corresponds to a haemoglobin concentration of 10 g%. This value should be compared with the direct measurement of haemoglobin and if the two values differ by more than 3 g%, this suggests an error in the acid–base determination.

Since so much can be read, the Siggaard-Andersen chart is naturally rather complicated in appearance. However, it is easy to use with practice. Consider a blood sample from a patient with compensated respiratory acidosis. The blood yielded the following two points:

Blood values	A	B
	at the *in vivo*	after equilibration
	P_{CO_2}	with $P_{CO_2} = 40$ mmHg
P_{CO_2}	71 mmHg	40 mmHg
pH	7.34	7.54

On the Siggaard-Andersen chart of Figure 4.10, these readings give points A and B which are then joined by a straight line. This line gives:

Base excess	10 mM
Buffer base	56 mM.

For this patient, the normal buffer base is $56 - 10 = 46$ mM. On the full Siggaard-Andersen chart of Figure 4.9, the haemoglobin value is read as 10 g%.

REVIEW OF METHODS OF PLOTTING ACID–BASE STATUS

The variables in the Henderson-Hasselbalch equation are $[H^+]$, $[HCO_3^-]$ and $[CO_2]$. Any pair of these can be used to plot graphs representing acid–base status; for each pair, each member can be plotted linearly or logarithmically. This book has explored two methods of representing acid–base status. In the earlier chapters, bicarbonate concentration was plotted as a function of the partial pressure of carbon dioxide, with both on linear scales. This was chosen for the following reasons.

The partial pressure of carbon dioxide is the factor which we deliberately vary in investigating the acid–base behaviour of blood. It is conventional to plot this variable on the x-axis. For following through the behaviour of a fluid exposed to different partial pressures of carbon dioxide, it is easy to visualize these changes as changes in the x-coordinates of points representing the fluid. For instance, in following through what happens if the P_{CO_2} is doubled, it is obvious where to move on the x-axis. This is an important feature in introducing the student to acid–base physiology.

Iso-pH lines are straight lines through the origin; it is easy to understand the construction of these lines and this provides a ready method of reading off the pH value corresponding to any point on the plot.

As explained in Chapter 6, the blood line on this plot is closely related to the carbon dioxide dissociation curve of the blood, a fact which allows acid–base physiology to be related directly to the carriage of carbon dioxide in the blood and thence to respiratory physiology.

The principal disadvantage of the representation is that the blood line is curved, not straight. A secondary disadvantage is that pH, the factor of prime importance in acid–base physiology, does not appear as one of the axes; this is only a minor problem, since the construction of iso-pH lines is so easy. The curvilinear nature of the blood line becomes an important problem when one arrives at the situation of wishing to use the display as a nomogram; for nomograms it is almost essential to be able to read off values with a ruler. The approach in this book has been to explain the basic physical chemistry of acid–base physiology with the help of the representation which was intuitively the easiest to understand and then, with this information assimilated by the reader, to move on to the logarithmic representation of the Siggaard-Andersen nomogram.

Other pairs of variables have been chosen by different authors. It is an advantage if lines representing constant values of the third variable, the one not represented on the x and y-axes, are straight lines, as is the case for iso-pH lines when $[HCO_3^-]$ is plotted as a function of $[CO_2]$. If $[H^+]$ and $[HCO_3^-]$ are plotted linearly against each other then iso-$[CO_2]$ lines are straight, whereas if $[CO_2]$ and $[H^+]$ are the variables iso-$[HCO_3^-]$ lines are hyperbolae. If both variables are plotted logarithmically, then constant values of the third variable are straight lines in all cases. If one variable is plotted linearly and the other logarithmically, then constant values of the third variable are curved. This is the approach of Davenport (1974) in his classical and masterly exposition of the subject; he plots $[HCO_3^-]$ as a function of pH. Here the P_{CO_2} isobars are curved. This representation has the advantage of displaying the buffering behaviour of a fluid as a titration curve. For fluids such as plasma and blood *in vitro*, the relationships are close to being straight lines. Davenport builds up the subject in these terms.

APPENDIX 4

A.1 Measurement of total buffer base

Various methods have been used. One method is to measure chemically the total concentration of cations (sodium, potassium etc.) and the total concentration of 'fixed' anions, so called because their concentrations are unaffected by changes in pH. The principal fixed anion is chloride, but others present in lower concentration include sulphate, nitrate etc. The difference between the total cation concentration and the fixed anion concentration is the buffer base (Singer and Hastings, 1948). Direct chemical analysis for the purpose of measuring total buffer base is complicated and is not performed routinely. The Siggaard-Andersen nomogram gives the total buffer base as one of the parameters which can be read off automatically when the line representing a patient's blood is plotted. The parameter is readily available to the clinician at no extra laboratory effort.

A.2 $\log [CO_2]$ as a function of pH for a constant value of $[HCO_3{}^-]$

$$\begin{aligned} pH &= pK + \log \frac{[HCO_3{}^-]}{[CO_2]} \\ &= pK + \log[HCO_3{}^-] - \log[CO_2] \\ \Rightarrow \log[CO_2] &= -pH + (pK + \log[HCO_3{}^-]) \end{aligned}$$

For a given value of $[HCO_3{}^-]$, the value within the brackets $(pK + \log[HCO_3{}^-])$ is a constant. So $\log [CO_2]$ as a function of pH is a straight line with a slope of -1.

A.1 Measurement of total linear heat

Acids and bases in the body | 5

SOURCES OF ACID OR ALKALI

By far the greatest production of chemical which threatens acid–base homeo-stasis is that of carbon dioxide in metabolism. Quantitatively, this amounts to 25 Moles of acid per day. This acid is blown off by the lungs (Table 5.1).

The food ingested by mouth in everyday life is usually neutral but it yields non-respiratory acids and alkalis as a result of metabolism. In normal circum-stances it is unusual for the production of acid exactly to balance that of alkali. The one which is produced to excess is excreted in the urine.

A normal mixed meat and vegetable diet yields an excess of acid. The main metabolic source of hydrogen ions is from meat; oxidation of sulphur-contain-ing amino acids gives sulphuric acid. Some phosphoric acid is also produced. As a result, the urine is usually acid. The daily load of acid from non-volatile sources is typically 70 mMoles which corresponds to 50% of the total acid held in all the buffer acids in the body. This is a very high turnover rate, and compares with values of 5% for sodium and 2.5% for potassium. Vegetables alone yield a preponderance of alkali so, as a result, the urine of vegetarian animals including vegetarian humans is alkaline. The homeostatic responses to acid or alkaline loads considered in previous chapters are being called into operation in health to keep the hydrogen ion concentration constant; they are not reserved as responses only to pathological conditions.

During starvation, body fats replace dietary carbohydrates as a major source of energy. Metabolism of these fats yields organic acids such as acetoacetic acid. This causes a mild metabolic acidosis. In strenuous exercise, particularly in a trained athlete, fats are preferentially metabolized; also, in exercise sufficiently strenuous to exceed the anaerobic threshold, lactic acid is released into the extracellular fluid in all subjects, both the trained athlete and the normal person. Fever and trauma such as extensive burns or major surgery result in tissue breakdown; this catabolism results in the release of organic acids into the extracellular fluid.

Table 5.1   Production of acid in the body

CO_2 from metabolism	Excreted via lungs	25 Moles per day
Meat-sulphur-containing amino acids yield sulphuric acid	Excreted via the kidneys	70 mMoles per day
Vegetables yield alkali		

BUFFERING OF AN ACID LOAD

The relative contributions of the bicarbonate and non-bicarbonate buffer systems depend on the nature of the acid load. If the load is due to carbon dioxide retention, the hydrogen ions produced by the addition of carbon dioxide to the blood cannot be buffered by the bicarbonate system. The reason is evident from examination of the chemical reaction:

$$CO_2 + H_2O = H^+ + HCO_3^-$$

The primary disturbance is a rise in concentration of carbon dioxide. This drives the chemical reaction to the right which is the origin of the rise in hydrogen ion concentration. The bicarbonate concentration also rises. By the law of mass action, the rise in $[CO_2]$ equals the rise in $[H^+]$ times the rise in $[HCO_3^-]$. The rise in CO_2 concentration is always proportionally greater than the rise in hydrogen ion concentration alone, so that this latter rise will not drive the equation back to the left, as would be required for buffer action. An intuitive way of looking at this situation is that the rise in $[CO_2]$ drives the chemical reaction to the right; buffering would require the reaction to be driven to the left. The reaction cannot be simultaneously driven in opposite directions. Bicarbonate cannot act as a buffer because it is a component of the carbon dioxide bicarbonate system. This is a particular case of the general principle that a buffer pair cannot buffer one of its own components.

Table 5.2   Whole-blood buffering of an acid load

Nature of acid load	Buffering		
CO_2	Non-bicarbonate buffers	100%	
Fixed acid	Bicarbonate $\begin{cases} \text{Plasma} \\ \text{Erythrocytes} \end{cases}$	35% 18%	} 53%
	Haemoglobin Plasma proteins Phosphate	35% 7% 5%	} 47%

As a consequence of the foregoing, all the buffering of a carbon dioxide load is due to non-bicarbonate buffer (Table 5.2). By contrast, a fixed acid is buffered both by the bicarbonate and the non-bicarbonate buffer systems; here bicarbonate provides quantitatively the greatest buffering, with plasma bicarbonate contributing 35% and bicarbonate in the erythrocytes contributing 18%. The proteins are next in line, with haemoglobin contributing 35% and the plasma proteins 7%. Phosphates contribute 5%. Together, bicarbonate in erythrocytes and plasma contributes over one half of the buffering power whilst non-bicarbonate buffer base contributes less than one half.

ACID–BASE PHYSIOLOGY IN THE CONTEXT OF BODY FLUIDS

So far in this book, the emphasis has been on the chemistry of acid–base balance. We now proceed to consider the relationship of the concentrations of the chemicals important in the context of acid–base balance to the composition of plasma. In the wider context of the physiology of body fluids such as extracellular and intracellular fluids, there are two constraints which are of importance. The first constraint is physico-chemical and applies to any aqueous fluid. It is called the 'principle of electroneutrality', which states that, in an aqueous solution, the number of positive charges equals the number of negative charges. For reference, this important principle is included in Table 5.3.

The second constraint is physiological and arises from the fact that the membranes of animal cells are frail and can withstand no pressure difference across them. Consequently, if there exists any difference in osmotic pressure between intra and extracellular fluids, there will be a net movement of water across the cell membrane until the osmotic gradient is cancelled. This net movement of water causes the cell to swell or shrink. At equilibrium, i.e. if the cell volume is not changing, the osmotic pressures of intra- and extracellular fluids are equal. These two different constraints give rise to two different units for measuring concentrations.

This is illustrated by consideration of a solution of potassium sulphate K_2SO_4 at a concentration of 100 mM. As shown in Figure 5.1A, each molecule of the salt dissociates into two ions of potassium and one ion of sulphate. The concentrations of the ions displayed as bar graphs are of different heights.

Table 5.3

1. *The Principle of Electroneutrality*
   In a solution, the number of positive charges equals the number of negative charges.

2. *Osmotic pressure*
   At equilibrium (i.e. when the volume of a cell is not changing), the osmotic pressures of intra- and extracellular fluids are equal.

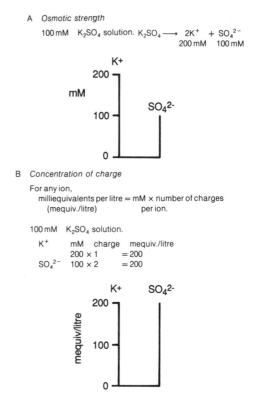

A *Osmotic strength*

100 mM K_2SO_4 solution. $K_2SO_4 \longrightarrow 2K^+ + SO_4^{2-}$
                                          200 mM   100 mM

B *Concentration of charge*

For any ion,
  milliequivalents per litre = mM × number of charges
    (mequiv./litre)   per ion.

100 mM K_2SO_4 solution.

	mM	charge	mequiv./litre
K^+	200 × 1		= 200
SO_4^{2-}	100 × 2		= 200

Figure 5.1. This figure indicates the need for two different units for measuring concentration of particles (osmotic strength) and concentration of charge. A. mM is the unit for measuring concentration of particles. This is the appropriate unit for indicating such colligative properties as osmotic pressure; the concentration of the solute is 300 mosmoles per litre. B. mequiv. per litre is the appropriate unit for measuring concentration of charge. This unit of measurement is used to check that the total concentration of positive charge equals the total concentration of negative charge. This is shown by the equal heights of the bars representing cations and anions.

This is because sulphate is divalent; its concentration in mM is half that of potassium.

For graphic representation of the principle of electroneutrality, this bar graph in millimols is uninformative. A more appropriate unit of concentration, 'the concentration of charge', is indicated in Figure 5.1B. For any ion, the concentration of charge is millimolarity multiplied by the number of charges on each ion. It is called **milliequivalents per litre** and is written **mequiv./litre**. It is a frequently used unit in American textbooks. For the potassium sulphate solution, as regards potassium ions, 200 mM corresponds to 200 milliequivalents per litre because potassium is univalent (Figure 5.1B). For sulphate, 100 mM corresponds to 200 milliequivalents per litre because sulphate is

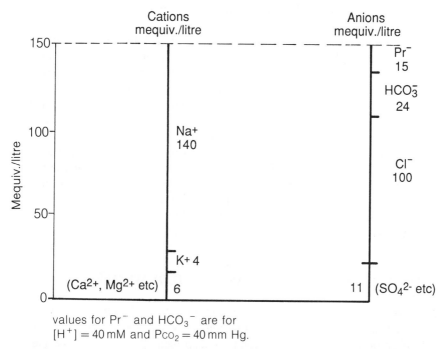

Plasma concentration

Figure 5.2. The composition of electrolytes in plasma. The unit of measurement is mequiv. per litre; the total concentration of charge contributed by cations equals that contributed by anions. The concentrations of protein anion and of bicarbonate are for blood with a P_{CO_2} of 40 mmHg and a $[H^+]$ of 40 nM.

divalent. Shown on a bar graph in Figure 5.1B, this displays electroneutrality as equality in the heights of cations and anions, when concentration is expressed in milliequivalents per litre.

COMPOSITION OF PLASMA

Units of milliequivalents per litre are used here to plot the composition of plasma in Figure 5.2 because the principle of electroneutrality is of interest; it demands that the sum of cations in mequiv./litre equals the sum of anions in the same units. Cations and anions are represented as separate bars and the sums of the two are therefore precisely equal in height, reflecting the principle of electroneutrality. The total for each bar is about 150 mequiv./litre. The principal cation quantitatively is sodium, at around 140 mequiv./litre. Potassium is around 4 mequiv./litre and the remainder, magnesium, calcium etc., give the balance of 6 mequiv./litre. The only cations of significance in acid–base physiology are hydrogen ions; their concentration is so small that no attempt is made to graph them in Figure 5.2.

Next to be considered are the anions. Since, in acid–base physiology, buffer base is of prime importance, it is sited at the top of the bar graph. Protein buffer base at 15 mequiv./litre and bicarbonate at 24 mequiv./litre are shown. These values for Pr^- and HCO_3^- correspond to a normal $[H^+]$ of 40 nM (pH 7.4) and a normal arterial P_{CO_2} of 40 mmHg. Next chloride, the principal anion, is shown with a concentration of around 100 mequiv./litre. The remainder, comprising sulphate etc., add up to 11 mequiv./litre.

A change in P_{CO_2} affects the relative amounts of bicarbonate and non-bicarbonate buffer base but the total remains constant. Changes in the partial pressure would not be expected to cause any change in the concentrations of ions such as chloride, because they do not take part in acid–base reactions. However, the reader should be forewarned that this simplistic conclusion is wrong, as will become clear later in this chapter.

TOTAL BUFFER BASE OF PLASMA AND OF WHOLE BLOOD

The value given for the total buffer base for plasma (taken from Figure 5.2) is $(15 + 24)$, i.e. typically 39 mM whereas that for whole blood is typically 48 mM. Whole blood is a better buffer than plasma, principally because of the haemoglobin contribution to buffering, which is why it has a larger concentration of total buffer base. It has already been noted that anaemic blood has a reduced amount of buffer base, reflecting the poorer buffering capacity by comparison with normal blood.

THE CHARGE ON ONE ION OF PROTEIN BUFFER BASE

With this insight into the difference between osmotic strength (concentration of particles) and electrical strength (concentration of charge), we can now review the status of protein buffer base. Until now, non-bicarbonate buffer base has been represented as Pr^-, a single negative charge per particle. This was a shorthand; in fact each protein particle carries on average 15 negative charges (Table 5.4). Strictly, it should have been written Pr^{15-}; the concentration of 15 mequiv./litre, shown in Figure 5.3, corresponds to a millimolarity of around 1 mM. Osmotic pressure depends on the concentration of particles and is uninfluenced by the charge on each particle. Hence the osmotic pressure exerted by the plasma proteins is to be calculated from the osmolarity, which is 1 mM. This gives an osmotic pressure due to the proteins of 25 mmHg.

In a quantitative approach to acid–base balance, it is the charge on buffer base which is of interest. Fifteen hydrogen ions combine with one protein ion

Table 5.4

Protein anion
This carries an average of 15 negative charges per ion

i.e. Pr^{15-}

In plasma, the concentration is:

$15\,mequiv./l = 1\,mM/litre.$

Protein exerts an osmotic pressure in plasma of about 25 mmHg.
Formation of protein buffer acid:

$$15H^+ + Pr^{15-} \rightarrow H_{15}Pr$$

to yield protein buffer acid (Table 5.4). Indeed, all the values previously given for buffer base and base excess should be in units of milliequivalents per litre. The total buffer base is actually normally 48 milliequivalents per litre, not 48 mM as quoted earlier in this book. This abuse of usage occurs because equivalents are no longer used in the teaching of physical chemistry in Britain. It introduces no anomalies so long as acid–base physiology is being considered in isolation from the wider context of the physiology of body fluids. As illustrated above, the anomalies arise when other physiological attributes are considered, e.g. the colloid osmotic pressure exerted by the plasma proteins as an aspect of their physiological importance different from the buffering of acids and alkalis.

THE BLOOD AS A BUFFER

In a solution in which there is a single buffer pair, the buffering characteristics can be encapsulated in a few parameters, such as the pK of the buffer and its buffer value at the pK. For a buffer pair such as NH_4^+/NH_3, these two parameters completely describe its behaviour as a buffer and its buffer curve can be forecast by appropriate scaling in Figure 1.3A. When more than one buffer grouping occurs on the same chemical, then the titration curve becomes more complicated; an example is afforded by phosphoric acid H_3PO_4 and its conjugate bases. Because it can donate three hydrogen ions per molecule, it has three pK values. The titration curve therefore has three points of inflection (Figure 5.3.). Although only one of the three pK values lies within the physiological range, the titration curve illustrates how several pK values tend to flatten out the titration curve over a wide range of pH values.

Blood is far more complicated than phosphate as a buffer system. There are many different chemicals contributing to buffering. Each haemoglobin molecule and each plasma protein molecule has many groups which donate and accept hydrogen ions and which have pK values in the physiological range. The result

$$H_3PO_4 \rightleftharpoons H^+ + H_2PO_4^- \rightleftharpoons 2H^+ + HPO_4^{2-} \rightleftharpoons 3H^+ + PO_4^{3-}$$

pK values   2.0   6.8   12.4

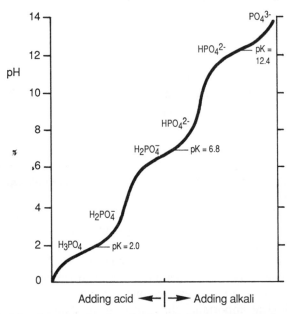

Figure 5.3. Titration curve for phosphoric acid. As shown by the chemical dissociation reactions at the top of the figure, one molecule of undissociated acid yields three hydrogen ions; correspondingly there are three pK values. Only the pK at 6.8 is near to the physiological range; the reason for showing the whole titration curve is to illustrate the flattening effect by comparison with the curves, such as that for CO_2 and HCO_3^- shown in Figure 1.3A, for an acid which yields only one hydrogen ion.

is that the titration curve is fairly flat over the physiological range of pH values; as the pH varies, one buffer grouping hands over to another. Other groups which accept or donate protons have pK values far from the physiological range of pH values. In experimentally determined titration curves, the pH can be pushed far beyond the limits compatible with life and contribution to buffering is made by chemical groupings which never have a chance to operate in life. It is analogous to peeling an onion and revealing layers (buffer groups) never normally exposed.

THE HENDERSON-HASSELBALCH EQUATION EXPRESSED IN TERMS OF CARBONIC ACID

The chemical equation which is the basis of acid base balance, i.e. the combination of carbon dioxide and water to yield hydrogen ions and bicarbonate ions, is

Table 5.5

A.	$CO_2 + H_2O \rightleftharpoons H^+ + HCO_3^-$
B.	$CO_2 + H_2O \rightleftharpoons H_2CO_3 \rightleftharpoons H^+ + HCO_3^-$

Typical concentrations (mM): 1.2, 0.004, 0.00004, 24

C. $pH = pK + \log \dfrac{[HCO_3^-]}{[H_2CO_3]}$ $pK = 3.6$

D. $pH = pK' + \log \dfrac{[HCO_3^-]}{[CO_2]}$ $pK' = 6.1$

now reviewed. So far in this book, the reaction has been assumed to occur in the form shown in Table 5.5A. There is doubt as to whether, in the tissues and in the lungs, the reaction proceeds directly, as this equation implies, or whether it proceeds via carbonic acid (Table 5.5 reaction B). The overall result is the same and, in a solution at equilibrium, all five species CO_2, H_2O, H_2CO_3, H^+, and HCO_3^- are present. The equilibrium constants are such that the amounts of chemicals are as shown in Table 5.5B; the concentration of carbonic acid is a small fraction of the concentration of carbon dioxide, when the quantities are expressed in mM.

For any weak acid, the chemical reaction of its dissociation into hydrogen ions and base is written:

$$HA = H^+ + A^-$$

In conformity with this convention, the basic chemical reaction in acid–base physiology is to be written:

$$H_2CO_3 = H^+ + HCO_3^-$$

This gives the Henderson-Hasselbalch equation in the form shown in Table 5.5C. Since the concentration of H_2CO_3 is so much less than that of CO_2, the pK is far below the physiological pH; the pK for this reaction is actually 3.6. More commonly, the chemical reaction:

$$CO_2 + H_2O = H^+ + HCO_3^-$$

is used as the basis of acid–base physiology. In this form, a new pK value is needed to allow for the relatively high concentrations of CO_2 compared with concentrations of H_2CO_3. To signify this difference, the pK is written with a prime, thus: pK'. This leads to the Henderson-Hasselbalch equation in the form shown in Table 5.5D, the pK' taking the value of 6.1, a figure familiar to the reader. Both these forms of the Henderson-Hasselbalch equation are equally valid (appendix).

THE COMPLEX NATURE OF BLOOD AND ITS ACID–BASE IMPLICATIONS

In all consideration of acid–base status to date, blood has been regarded as a simple fluid consisting of one phase only. The fact is that blood consists of two discrete phases, plasma and erythrocytes, shown diagrammatically in Figure 5.4. This introduces complications which must now be considered. When measurements are made on blood to determine its acid–base status, the partial pressure of carbon dioxide is measured with a blood gas meter and the pH is measured with a glass electrode. The $[HCO_3{}^-]$ is calculated from the Henderson-Hasselbalch equation. The pH electrode is in the plasma and thus samples the hydrogen ion concentration of the plasma, but the composition of the blood is not uniform and the standard analyses yield no direct information on the composition of the intracellular fluid of the erythrocytes. The exact relationships between intracellular and extracellular fluids are not known with precision. Thus, having developed the subject of acid–base physiology, we now realize that the acid–base status measurements which we have been considering throughout are measurements on the plasma phase of blood.

At this stage, two definitions are introduced:

Separated plasma: this is plasma separated from the red cells. Blood is centrifuged and the plasma is then sucked off and studied in isolation.

True plasma: this is plasma in equilibrium with its red cells. This nomenclature is intended to draw attention to the fact that the blood analysis yields concentrations in the plasma phase. To be accurate, whenever graphs have been drawn and have been said to indicate the behaviour of blood, in fact they have indicated the behaviour of true plasma, the plasma moiety of the whole blood.

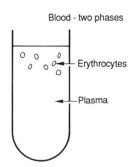

Blood - two phases

Measure Pco_2 with blood gas meter pH with glass electrode.
Calculate $[HCO_3{}^-]$ by Henderson-Hasselbalch equation.
It is the composition of the plasma which is measured.

Definitions

Separated plasma is plasma separated from the erythrocytes
True plasma is plasma in equilibrium with its erythrocytes

Figure 5.4. Blood consists of two phases, plasma and erythrocytes.

If the erythrocyte membrane is lysed, for instance by adding a small amount of saponifying agent (detergent), then the system becomes a single phase and its behaviour is much easier to analyse. The behaviour is, however, quite different from that of normal blood and we do not study it because it is of no relevance to physiology. The fact that the blood has two distinct phases makes it a very complicated system. This complication is compounded by the fact that we can measure the concentrations of chemicals in only one of these phases, the plasma.

THE 'CHLORIDE SHIFT'

An example of the complications caused by the fact that blood consists of two phases is the effect that changing the P_{CO_2} has on the concentrations of anions in these two fluids. For separated plasma, increasing the P_{CO_2} will lead to an increase in $[HCO_3{}^-]$ and a parallel decrease in $[Pr^-]$. The chloride concentration will remain unaltered, because chloride ions, being the conjugate base of hydrochloric acid which is a very strong acid, are not directly involved in

Chloride shift

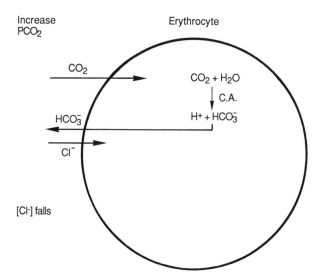

Figure 5.5. The chloride shift. An increase in P_{CO_2} to which blood is exposed results in a net entry into erythrocytes of CO_2; within the erythrocyte, the hydrolysis of CO_2 is catalysed by carbonic anhydrase (C.A.). The rise in concentration of bicarbonate within the erythrocyte provides an outwardly-directed concentration gradient and a consequent efflux of bicarbonate ions. Electrical neutrality across the membrane is preserved by an influx of chloride ions and the concentration of chloride in the plasma falls.

acid–base chemistry. In blood, however, increasing the P_{CO_2} with which blood is equilibrated results in the so-called 'chloride shift' (Figure 5.5). Most of the carbon dioxide taken up by the blood diffuses immediately into the erythrocytes, where it is hydrated with the assistance of carbonic anhydrase, present in the erythrocyte cytoplasm but not in plasma. The hydrogen ions released by the hydration are buffered by haemoglobin. The intracellular bicarbonate concentration rises, thereby creating an outward gradient for bicarbonate movement. The movement of bicarbonate across the erythrocyte membrane is principally by means of an exchange mechanism which carries chloride in the opposite direction. An outward gradient for bicarbonate results in bicarbonate efflux accompanied by chloride influx, this coupled exchange ensuring electrical balance. The transmembranal movement of chloride ions is the 'chloride shift'. The total amount of bicarbonate in blood is unchanged by this shift but the distribution between erythrocyte cytoplasm and plasma is altered. The same is true of chloride. Measurement of the chloride concentration in true plasma thus reveals a fall when the P_{CO_2} of gas with which the blood is in equilibrium rises. This is despite the fact that chloride takes no direct part in buffering reactions.

BLOOD IN THE BODY

The next simplification which has been made so far in this book is to consider the acid–base status of the blood as if it were held in a glass vessel instead of circulating in the blood vessels of the body. *In vivo*, the plasma interfaces not only with the red cells (as for blood *in vitro*) but also with the interstitial fluid, across the capillary membrane. All small ions and molecules exchange readily between plasma and interstitial fluid. The magnitude of the interstitial fluid is 11 litres, while the plasma is three litres, so the interstitial fluid acts as a huge reserve for interface with the plasma (Figure 5.6A). Interstitial fluid is virtually protein-free and is therefore inferior as a buffer to blood or even to plasma. Hence the presence of interstitial fluid makes blood *in vivo* behave as if it were much more weakly buffered; this is shown in Figure 5.6B. However, the cells of the body and the bone represent a huge reservoir of buffering power and act as a buffer back-up to the interstitial fluid in a manner analogous to that in which the erythrocytes act as a buffer back-up to plasma; this effect is also indicated in Figure 5.6B. So these two *in vivo* effects act in opposite directions, tending to cancel each other.

The response to the whole body to respiratory and metabolic stress has been studied in human volunteers. It turns out that the *in vivo* blood line is very close to the *in vitro* blood line. Some individuals show a steeper relationship of $[HCO_3^-]$ as a function of P_{CO_2}; others show a less steep relationship. It is fortunate that the behaviour of blood *in vitro* approximates sufficiently closely to that of the whole body for *in vitro* behaviour to be useful clinically.

A

Blood in the body

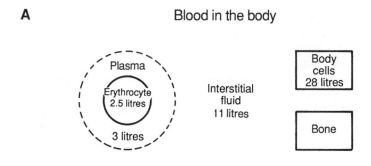

B

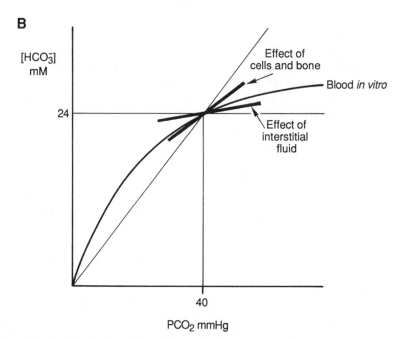

PCO$_2$ mmHg

Figure 5.6. A. The plasma of blood in the body interfaces with interstitial fluid which in turn interfaces with the body cells and with bone. B. The effect of the interstitial fluid is to reduce the buffering power of blood *in vivo* whilst the effect of the buffering capacity of intracellular fluid and of bone is to increase the buffering capacity.

However, it is important to remember that the disturbance is assessed from the chemistry of a sample of patient's arterial blood *in vitro* and not from the biochemistry of the whole patient. For any one patient, the inaccuracy introduced is unknown. To this extent, measures such as the base excess are empirical and the doctor must be aware of these limitations.

A particular case where the measurements tend to be misleading is in acute severe respiratory acidosis. Because of the interchange of ions between blood and interstitial fluid, there is relatively frequently a base excess of 10 mM or more in the absence of compensation. This is because base excess is measured by studying the chemistry of blood *in vitro*, equilibrating the blood with Pco_2 of 40 mmHg and back-titrating to a pH of 7.4. If blood could be subjected to the same manoeuvres *in vivo*, the results would be slightly different.

Most clinicians, whilst being wary of cases such as this, have sufficient faith in the magnitude of the base excess measured from a sample of arterial blood *in vitro* to use it in conducting their treatment. The rationale of their approach is as follows.

TREATMENT OF ACID–BASE DISORDERS

Disorders of acid–base balance are usually a consequence of disease such as diabetes mellitus. The main thrust of treatment is reversing the primary disease process and then the regulatory mechanisms of the body automatically reverse the acid–base abnormalities. There are, however, conditions in which the acid–base disorder must be treated directly because otherwise the patient dies of the acid–base upset before the primary illness can be treated. Acidaemia in particular needs treatment. In most pathological conditions involving metabolic disorders of acid–base physiology, the result is acidaemia, alkalaemia being much less common.

Metabolic acidosis occurs as a consequence of many pathological conditions (Table 5.6). In addition to diabetes mellitus another example is cardiac failure, for instance following a severe coronary thrombosis. Here, tissue oxygenation

Table 5.6

A. *Some causes of metabolic acidosis*
   Diabetes mellitus
   Cardiac or ventilatory failure: inadequate oxygenation of the tissues:
      anaerobic metabolism: release of lactic acid.
   Renal failure: retention of sulphuric acid.

B. *Effects of acidaemia*
   Coma
   Depression of cardiac contractility
   Depression of sympathetic and enhancement of parasympathetic effects on
   the heart.

C. *Treatment*
   Bicarbonate space is about 1/3 body weight. Total base excess of the whole
   body (mMol)

$$= \frac{\text{body weight (kg)}}{3} \times \text{Base excess (mM)}.$$

   Give half of the estimated total base deficit.

---

is compromised and the tissues respire anaerobically, releasing lactic acid. Respiratory failure has the same effect. In renal failure, the failure to excrete sulphuric acid from metabolism leads to metabolic acidosis.

A pH of below 7.25 is associated with mental disturbances and coma. Administration of bicarbonate brings the patient back to consciousness, confirming that the high hydrogen ion concentration is the prime cause of the coma. Acidaemia depresses the heart, both directly by depressing cardiac contractility and indirectly via the cardiac innervation, by depressing sympathetic and enhancing parasympathetic effects on the heart.

If acidaemia is itself life-threatening to a patient, an intravenous infusion of sodium bicarbonate or sodium lactate solution may be given. If the urgency is not so great, a sodium citrate and citric acid mixture by mouth may be given.

The magnitude of the base excess may be used as an indicator of the amount of bicarbonate which should be administered. As already indicated, there are several complicating factors which must be borne in mind when using this approach. The base excess indicates how much bicarbonate to add to blood at a P_{CO_2} of 40 mmHg in order to return the pH to 7.4; if the respiratory system is malfunctioning the abnormal P_{CO_2} must be taken into account. The estimate of base excess is made on the blood, but added bicarbonate interacts with many buffer systems in the body, of which those in blood are a subset. The fact that added bicarbonate equilibrates throughout the body water is incorporated in the calculations as follows.

The intracellular concentration of bicarbonate is typically 10 mM compared with the extracellular concentration of 24 mM. Bicarbonate added to the body distributes rapidly throughout the extracellular fluid, which is about one third of the total body water, and much more slowly through the intracellular water, which is about two thirds of the total body water. In order to calculate how much bicarbonate to administer, one needs to know both the increase in extracellular bicarbonate concentration which is required and also the volume within which administered bicarbonate will be distributed. This volume is called the 'bicarbonate space'. It is less than the total body water because of the lower intracellular than extracellular concentration of bicarbonate. Directly estimated, the bicarbonate space is approximately one third of the total body weight. The total base excess of the body is estimated (Ledingham and McKay, 1988, p. 50) as 0.3 times the body weight times the base excess in mM (Table 5.6C). Half the estimated total base deficit is administered as an infusion of $NaHCO_3$. The infusion is given slowly; it has been found by experience that bicarbonate is more effective given in this way rather than given rapidly, although the reason for this difference is unknown. Time is allowed for the bicarbonate to have its effect and then the patient is re-assessed. Further doses of bicarbonate are administered as necessary. In this way, the physician back-titrates the whole patient towards a normal blood biochemistry. The base excess is thus used as a rough guide to the amount of bicarbonate to be given to a patient. It is unnecessary, and potentially hazardous, to correct the acidosis completely.

One reason is that the administration of sodium ions with bicarbonate expands the extracellular space, which may carry the patient into cardiac failure.

INTRACELLULAR HYDROGEN ION CONCENTRATION

Until now, we have mainly considered extracellular hydrogen ion concentration. The usefulness to the organism of a stable extracellular fluid is so that intracellular biochemical processes occur at constant hydrogen ion concentration. Enzymic activity is extremely sensitive to small changes in pH; for example the enzyme phosphofructokinase increases in activity almost 20-fold if the pH changes from 7.1 to 7.2.

The intracellular pH is about 0.2 units more acid than the extracellular pH. This corresponds to an equilibrium potential for H^+ across the cell membrane of about $-20\,mV$. Bicarbonate and hydroxide ions also have equilibrium potentials of this value. The membrane potential is typically much more negative than this. Although these ions do not permeate the cell membrane readily, they are able to cross slowly and such passive movement results in a net transfer (H^+ inwards, OH^- and HCO_3^- outwards). There must therefore be active transfer across the cell membrane of one or more of these ions in the opposite direction in order to maintain the observed concentration gradients.

COMPARISON OF EXTRA- AND INTRACELLULAR HYDROGEN ION CONCENTRATION IN DISORDERS OF ACID–BASE PHYSIOLOGY

In respiratory disturbances of acid–base physiology, the intra- and extracellular $[H^+]$ move together, since carbon dioxide readily permeates cell membranes. In non-respiratory disturbances, the intra- and extracellular $[H^+]$ may diverge in extent and even in direction, as was demonstrated in gastric alkalosis, described in Chapter 3; to some degree this divergence is due to the fact that charged particles such as HCO_3^- ions do not readily permeate the membranes of most cells.

HOMEOSTASIS

Maintenance of a constant composition of the extracellular fluid is important because it provides a stable environment for the cells of the body. This was first recognized by Claude Bernard, who described the extracellular fluid as the 'milieu interiéur' or the '**internal environment**'.

In very primitive organisms, cells are bathed in the sea in which they live. The cell environment is very variable and at the mercy of the tides and winds. As evolution proceeded, organisms walled off some of this watery environment and control mechanisms developed to hold its composition at a more constant level. This is the origin of the interstitial fluid which in evolutionary terms is the 'sea within us'. In higher animals, the interstitial fluid is called the internal environment. It is the environment of the body's cells. Examples of factors which are stabilized include osmolarity, temperature and the concentrations of hydrogen ions, individual electrolytes, organic chemicals such as glucose concentration and gases such as O_2 and CO_2.

The need for this stability is that the body is a highly complex biochemical machine depending on efficient operation in concert of enzymes. Each enzyme's activity depends on many factors. If the internal environment wanders in composition this harmonious action of enzymes breaks down. The cells of the mammalian brain in particular cannot tolerate more than small changes in temperature, osmotic pressure, glucose concentration or acidity. The tendency of the body to stabilize the composition of the extracellular fluid is called **homeostasis** which means 'staying similar'. Physiological mechanisms subserving this function are called 'homeostatic mechanisms'. The control of the composition of the internal environment had to be achieved before the higher animals could evolve. In the words of Joseph Barcroft (1938), 'To look for high intellectual development in a milieu whose properties have not become stabilized is to seek music amongst the crashings of a rudimentary wireless or the ripple patterns on the surface of the stormy Atlantic'.

In acid–base balance, the constancy of the hydrogen ion concentration of the extracellular fluid is the factor of interest. Chemical buffering of hydrogen ion concentration is the first line of defence against changes in pH. The intracellular fluid of the cells of the body generally provides a much greater buffering capacity than the extracellular fluid itself. For example, the buffering provided by the erythrocytes, as a back-up to the stabilization of the pH of the plasma, with its low buffering capacity, has previously been described. This is an example of the general rule that, when intracellular buffering contributes to stabilizing extracellular pH, the cell is giving away its own intracellular buffering to the extracellular fluid. The cell is contributing to the provision of a stable internal environment for other cells at the expense of its own intracellular pH stability. From the point of view of the cell providing buffer, this mechanism is counter-productive in that it lessens the buffering of intracellular pH.

The back-up provided by the erythrocytes is of this nature. For these particular cells, the change in intracellular pH is of little consequence because erythrocytes, being non-nucleated cells, have rather sparse intracellular enzyme-dependent mechanisms. The mechanism occurs not only with erythrocytes but also for all cells whose intracellular buffering capacity is greater than that of the extracellular fluid and whose membranes allow chemicals involved in acid–base reactions to cross. The cells other than erythrocytes are in a quite different situation and, for them, changes in intracellular pH may be catastrophic. It is useful in this context to differentiate between erythrocytes as non-nucleated cells and the rest of the cells in the body as the nucleated cells (Burton, 1992).

Some nucleated cells which are particularly sensitive to changes in pH have mechanisms for stabilizing their own intracellular pH by active pumping of acid or base between intra- and extracellular fluid across the cell membrane in the direction which stabilizes the intracellular pH. A complicated system of membrane pumps is involved in this regulation (Thomas, 1984) and the regulation is so effective in some cells that no change of intracellular pH can be detected as the PCO_2 is varied. Such mechanisms result in changes in extracellular composition in a direction to exacerbate the extracellular acid–base disturbance.

Consider for instance the response to hypercapnia of an organism with a preponderance of cells which regulate their own intracellular pH. The rise in intracellular P_{CO_2} causes the intracellular release of hydrogen ions. Maintenance of a constant intracellular $[H^+]$ requires a rise of intracellular $[HCO_3^-]$ in direct proportion to the rise in intracellular $[CO_2]$; the regulation of the cell involves inward transfer of $[HCO_3^-]$ and consequently a fall in extracellular $[HCO_3^-]$. This exacerbates extracellular acidosis. Direct evidence for this mechanism comes from experiments with the salamander where Heisler *et al.* (1982) have shown that the initial response to hypercapnia is a large fall in extracellular $[HCO_3^-]$ as cells regulate their own intracellular pH. In mammals, the response to hypercapnia is a rise in extracellular $[HCO_3^-]$, presumably because these animals have a preponderance of cells which do not regulate their own pH so effectively. The effect here is a cooperation amongst the various cells in the body. Those which do not need an exceptionally stable intracellular pH contribute to providing stability of the extracellular fluid for cells elsewhere in the body which do have this requirement.

This is an illustration of some general features of homeostasis. A homeostatic mechanism for stabilizing the concentration of a chemical (such as $[H^+]$) in the extracellular fluid must involve the excretion of the chemical into the outside world or absorption of the chemical from it, e.g. excretion of $[H^+]$ into the urine via the kidney, absorption of calcium from ingested food in the gut. Exchange with cell contents will not work, in the long term, because it results in a change in composition of the intracellular fluid and hence defective intracellular homeostasis. It is analogous to the situation of being in debt. You cannot, in the long term, redress the situation by transferring money from one banking account in your own name to another in the same name. You have to get more money into the accounts from the outside world.

There are exceptions to this rule, for instance for chemicals which the body can itself synthesize. If the kidney is excreting large quantities of acid, it can manufacture ammonia which is used to buffer the urinary hydrogen ions.

APPENDIX 5

A.1 Carbon dioxide and carbonic acid

The use of the prime in pK′. This is used as defined in the present text to indicate the pK value when the concentration of CO_2 is used rather than that of H_2CO_3 in the Henderson-Hasselbalch equation. Physical chemists use the prime for a different purpose. They use pK for the constant expressed in terms of concentration and pK′ for the constant expressed in terms of activities, i.e. as actually measured experimentally.

Confusion between the values of pK for CO_2 and H_2CO_3. Some texts which are otherwise excellent and authoritative confuse these two pK values and the chemical equilibrium which they are using; the reader who wishes to read further may be puzzled if not forewarned. Presumably the origin of the confusion is due to the fact that molecular CO_2 is not itself an acid; it is actually the concentration of CO_2 which is measured. Carbonic acid is correctly referred to as the conjugate acid of bicarbonate but the concentration of CO_2 is used instead of the concentration of H_2CO_3.

Oxygen carriage, respiration and acid–base physiology

6

SECTION 6.1 CARRIAGE BY THE BLOOD OF OXYGEN AND CARBON DIOXIDE

THE OXYGEN DISSOCIATION CURVE OF THE BLOOD

The oxygen dissociation curve is the relationship between the amount of oxygen held in the blood and the partial pressure of oxygen in the gas with which the blood is in equilibrium. (See section A.1 for a note on nomenclature.) In the physiological range of partial pressures of oxygen, most of the oxygen held in the blood is in combination with haemoglobin. There is a small amount of oxygen in physical solution; this amount is directly proportional to the partial pressure of the oxygen with which the blood is equilibrated. At a partial pressure of 100 mmHg, one litre of blood holds 3 ml oxygen in physical solution whereas it holds 200 ml oxygen in combination with haemoglobin.

In order to plot a point on the dissociation curve, a sample of blood is placed in a glass vessel called a tonometer, as shown in Figure 6.1A. With the taps open, a gas mixture containing a known partial pressure of oxygen is passed through the tonometer and then the taps are closed. The tonometer is rotated for half an hour in a water bath at 37°C, by which time the blood and gas are in equilibrium. The volume of gas is so much larger than the volume of blood that the partial pressure of oxygen in the gas is insignificantly altered by exchange of oxygen with the blood.

When mixing is adequate, a measured volume of blood is withdrawn through one of the tonometer's taps and all the oxygen is extracted from the blood. The amount of oxygen so yielded is measured and this gives a point on the graph. The procedure is repeated for a range of partial pressures of oxygen, each estimation giving another point on the graph. These points are joined to give the oxygen dissociation curve. A typical curve for blood taken from a healthy adult person is shown in Figure 6.1B. The x-coordinate is the partial

A

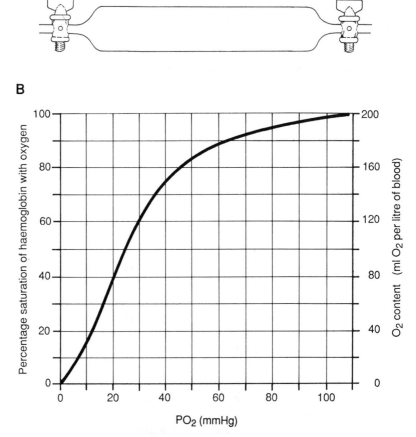

B

Figure 6.1. A. A tonometer, the glass vessel in which a sample of blood can be brought to equilibrium with gas of known composition. The volume of the tonometer is around 200 ml and the blood sample is 2 ml. This ensures that changes in the composition of the gas as a result of absorption by the blood or evolution from the blood are minimized. B. Oxygen dissociation curve of normal blood. The x-axis is the partial pressure of oxygen in the gas with which the blood is equilibrated. The y-axis is the oxygen content of the blood; on the left, this axis is calibrated as a percentage of the total oxygen-carrying capacity of the haemoglobin and on the right as the amount of oxygen, in units of ml oxygen per litre of blood.

pressure of oxygen. The y-coordinate can be labelled in two ways, as shown to the left and right of the plot. With high partial pressures of oxygen the haemoglobin is saturated with oxygen; the amount of oxygen held by the haemoglobin in this situation is called the 'oxygen-carrying capacity'. The amount of oxygen at lower partial pressures can be expressed as a percentage of total saturation; this is the left scale in Figure 6.1B. Alternatively, the actual amount of oxygen, in terms of ml O_2 per litre of blood, may be plotted and this is shown as the scale on the right. For a person suffering from anaemia, in whom the oxygen-carrying capacity of the blood is reduced, the values shown on the right would be scaled down by a constant; this constant is the concentration of haemoglobin in the anaemic person's blood divided by the normal concentration of haemoglobin.

The relationship is influenced by physiological variables other than the partial pressure of oxygen itself; the partial pressure of carbon dioxide is one important factor and the temperature is another. To plot a meaningful oxygen dissociation curve, it is therefore necessary to keep such factors constant. The gas mixtures used for equilibrating the blood samples at different values of oxygen partial pressure must all contain carbon dioxide at a given partial pressure (in Figure 6.1B, the P_{CO_2} was maintained at 40 mmHg). All estimations were made at 37°C.

Each molecule of haemoglobin has four binding sites for oxygen. As the oxygen partial pressure increases from zero, the uptake of oxygen by haemoglobin is initially small but, as a molecule of haemoglobin takes up the first molecule of oxygen, this changes the arrangement of protein subunits of the haemoglobin molecule in such a way that the affinity for the second molecule of oxygen is enhanced. This is why the slope of the dissociation curve becomes steeper. As the oxygen partial pressure rises further, the third and fourth binding sites on the haemoglobin molecule become occupied. After this stage, the haemoglobin molecule can accommodate no more oxygen. This is why the dissociation curve reaches a plateau; at this stage the haemoglobin is saturated with oxygen. Further increases in oxygen partial pressure will now fail to increase the amount of oxygen held by haemoglobin. As a result of these factors, the oxygen dissociation curve is S-shaped, alternatively called sigmoid.

TYPICAL VALUES FOR P_{CO_2} AND OXYGEN CONTENT IN ARTERIAL AND MIXED VENOUS BLOOD

Arterial blood in a normal person at rest breathing air at sea level contains typically 200 ml oxygen per litre and the partial pressure is close to 100 mmHg. For venous blood from different organs, the content and partial pressure of oxygen is variable because of the different usages and regional distributions of blood. To arrive at an average figure for the venous return from the body as a whole, it is necessary to take a sample of blood where all the different venous drainages have mixed; such blood can be obtained from a catheter lying in the right atrium. To achieve even better mixing, the catheter should ideally be advanced through the atrio-ventricular valve to sample blood from the right ventricle. Such blood typically contains 150 ml O_2 per litre of blood and the partial pressure is 40 mmHg.

THE BOHR EFFECT

Bohr found in 1910 that the position of the oxygen dissociation curve depended on the P_{CO_2} in the gas mixture with which the blood was equilibrated. This effect is now known as the **Bohr effect** and is shown in Figure 6.2. The oxygen-carrying capacity, i.e. the amount of oxygen contained in blood when it is fully saturated at high partial pressures of oxygen, is unaltered. For lower partial pressures of oxygen, for instance values found in mixed venous blood, an increase in the partial pressure of carbon dioxide shifts the oxygen dissociation curve to the right. This means that at a given P_{O_2} the amount of O_2 held in the blood is decreased when the P_{CO_2} is increased. As blood traverses the systemic capillaries in the body, the partial pressures of both oxygen and carbon dioxide are changing, with the oxygen partial pressure falling as oxygen is off-loaded into the tissues and the CO_2 partial pressure rising because of carbon dioxide yielded by tissue metabolism being taken into the blood. Because of this, the oxygen dissociation curves plotted in Figures 6.1B and 6.2 are purely theoretical curves. As the blood flows through the capillaries, the representation on a plot of oxygen content as a function of oxygen partial pressure moves from one of the dissociation curves of Figure 6.2 to another.

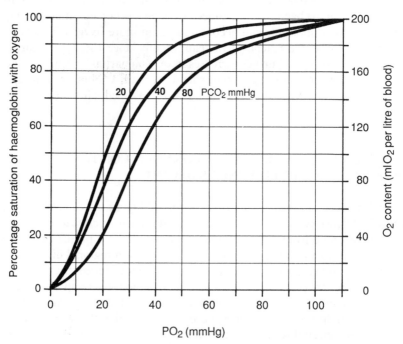

Figure 6.2. Oxygen dissociation curves at three different partial pressures of carbon dioxide. Increasing the P_{CO_2} shifts the curve to the right.

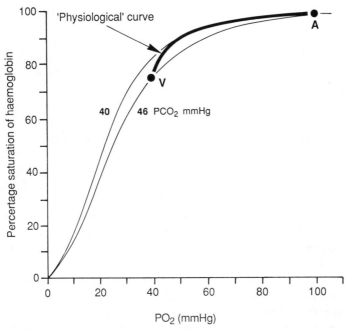

Figure 6.3. The 'physiological' oxygen dissociation curve for a normal person at rest; this is the curve representing the changes in P_{O_2} and oxygen content of blood as it passes through the systemic capillaries (the trajectory is from A to V) and through the pulmonary capillaries (from V to A). The physiological curve deviates from the curve plotted at a given P_{CO_2} because as the blood traverses the capillaries, both the P_{O_2} and the P_{CO_2} change. A and V show the P_{O_2} and content typical of arterial and mixed venous blood respectively. The figure also shows the oxygen dissociation curves for two values of P_{CO_2} typical of arterial and mixed venous blood (40 and 46 mmHg respectively). In this figure, the curves are not accurately drawn and the effect of increasing P_{CO_2} on the oxygen dissociation curve is exaggerated for clarity.

Figure 6.3 shows the form of this so-called 'physiological' dissociation curve which is steeper than the theoretical dissociation curves. Physiologically, this confers the advantage that oxygen is yielded by the blood to the tissues with a minimal change in oxygen partial pressure. The maintenance of adequate oxygen partial pressures in the tissues is essential for aerobic metabolism to be possible. Mitochondria need an oxygen partial pressure in excess of 1 mmHg; below this partial pressure, which is called the 'Pasteur point', aerobic metabolism ceases. In tissues such as muscle, energy can be supplied by anaerobic metabolism, although this is much less efficient than aerobic metabolism in terms of energy yielded per gram of glucose metabolized. In certain other tissues, notably brain tissue, energy is obtainable only by aerobic metabolism and if the oxygen partial pressure falls brain cells die.

The movement of O_2 from the blood to the mitochondria is by diffusion. Net movement occurs from regions where the partial pressure is high to regions where the partial pressure is low. The driving force for oxygen movement is the difference in partial pressures along this path. There is therefore a fall of oxygen partial pressure at every stage as the oxygen travels on its journey from the blood through the interstitial fluid across the cell membrane and intracellularly to the mitochondrial membrane. Together with the movement of oxygen from the air to alveoli, to pulmonary capillary blood and via the arterial blood to the capillary, this is known as the 'oxygen cascade'. The mean oxygen partial pressure at the level of the capillary wall is typically 43 mmHg, just over the value for the blood at the venous end. Despite this relatively high value, the decrease in partial pressure associated with diffusion of the oxygen from capillary to the intracellular compartment results in the partial pressure of oxygen near to mitochondria, even in tissues at rest metabolizing at a relatively slow rate, being close to 1 mmHg. Any fall in O_2 partial pressure at the capillary will result in a fall in Po_2 at the mitochondrial end to below the Pasteur point. In this context, the effect of carbon dioxide on the oxygen dissociation curve of the blood contributes to maintaining the partial pressure of oxygen in the tissues.

A rise in temperature also shifts the oxygen dissociation curve of the blood to the right. The physiological advantage conferred by this behaviour is that, in exercising muscle, the heat generated by the muscle raises the local temperature and helps in the off-loading of oxygen from the blood to the tissues. This is just one of many components of the adaptive responses called into play when a tissue metabolizes rapidly.

ACID–BASE BALANCE AND THE CARBON-DIOXIDE DISSOCIATION CURVE OF THE BLOOD

In plots of acid–base status, bicarbonate was displayed as a function of the partial pressure of carbon dioxide. One of the reasons for choosing this pair of variables is that the graph given for blood on the acid–base status plot is very similar to the carbon dioxide dissociation curve for the blood. The carbon dioxide dissociation curve is the relationship between the partial pressure with which blood is in equilibrium and the amount of carbon dioxide which the blood holds.

To determine the carbon dioxide dissociation curve, one proceeds as follows. A sample of blood is equilibrated with a gas mixture containing a known partial pressure of carbon dioxide; this value is plotted as the x-coordinate. A strong acid such as hydrochloric acid is then added and the blood is exposed to a vacuum to drive off all the carbon dioxide. The amount of carbon dioxide extracted is measured and plotted as the y-coordinate to yield one point on the graph. The procedure is repeated on samples of blood at various values of carbon dioxide partial pressure. At each Pco_2 the content is measured and the points on the graph are joined to give the dissociation curve (Figure 6.4).

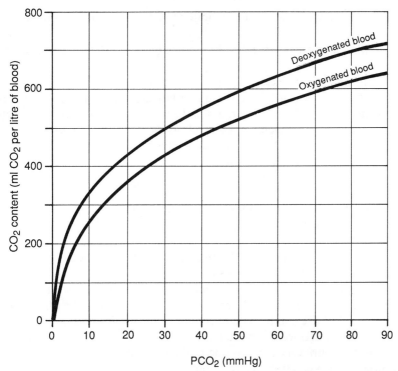

Figure 6.4. Carbon dioxide dissociation curves for oxygenated and deoxygenated blood.

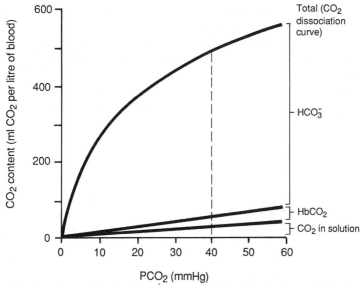

Figure 6.5. Graph showing the forms in which carbon dioxide is carried in the blood. The lowest graph is the amount of CO_2 in physical solution. The next is of CO_2 in physical solution plus carbamino haemoglobin. The top curve adds in the amount carried as bicarbonate.

The graphs for oxygenated and deoxygenated blood are different, as described later.

The carbon dioxide yielded by this method comes from various sources in the blood. As shown in Figure 6.5, quantitatively the most important is bicarbonate (24 mM). Next are carbamino haemoglobin $HbCO_2$ (a compound of carbon dioxide and haemoglobin) and carbon dioxide in physical solution. Each contributes about 1.2 mM. The contribution of carbonic acid is so small that it cannot be shown on this scale.

UPTAKE AND RELEASE IN CAPILLARIES OF CARBON DIOXIDE

Carbon dioxide diffuses from the site of production in metabolizing cells, through the interstitial fluid, across the wall of the systemic capillary and into the blood plasma. It readily diffuses across the membrane of the erythrocyte. When carbon dioxide enters the erythrocyte, bicarbonate and hydrogen ions are formed. The rise in intracellular concentration of bicarbonate leads to the chloride shift, described in Chapter 4. The hydration of carbon dioxide in the erythrocyte occurs very quickly, due to the presence of the enzyme **carbonic anhydrase**, which catalyses the reaction:

$$CO_2 + H_2O = H^+ + HCO_3^-$$

The speed of reaction within the erythrocytes is about 13 000 times faster than in the plasma. This accounts both for the rapid uptake of carbon dioxide by the blood during the short transit time (about one sec.) of blood through the systemic capillary and for the rapid release of carbon dioxide from the blood to the alveolar gas during the transit of the blood through the pulmonary capillaries. Without carbonic anhydrase, carbon dioxide partial pressures in the venous blood, and hence in the tissues would be much higher. Indeed when

Table 6.1 Carbon dioxide

	Total content (arterial blood)		A-V difference	
	Amount (ml CO_2 per litre of blood)	Percentage of total	Amount (ml CO_2 per litre of blood)	Percentage of total
HCO_3^-	450	90	35	70
$HbCO_2$	25	5	10	20
CO_2 in physical solution	25	5	5	10
Total	500		50	

drugs which inhibit carbonic anhydrase are administered for therapeutic purposes, partial pressures of carbon dioxide in the tissues do rise, with a resultant acidity.

On average in a resting person, each litre of blood takes up 50 ml of carbon dioxide. The forms in which it is carried are shown in Table 6.1. Deoxygenated haemoglobin forms carbamino haemoglobin more readily than does oxy-haemoglobin. Although in absolute terms the amount of carbon dioxide held in the blood in the form of carbamino haemoglobin is small (5% of the total), the arterio-venous difference is large (20% of the total arterio-venous difference, Table 6.1).

COMPARISON OF THE OXYGEN AND CARBON DIOXIDE DISSOCIATION CURVES OF BLOOD

Since carbon dioxide reacts with water to yield bicarbonate which is readily transported by the blood, no special carrier molecule, analogous to haemo-

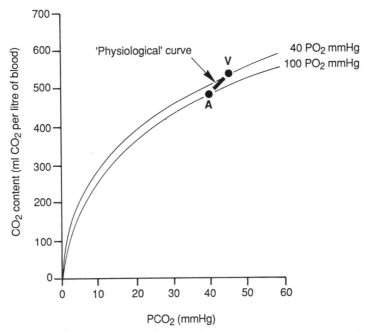

Figure 6.6. The 'physiological' carbon dioxide dissociation curve for a normal person at rest. The points A and V represent partial pressures and contents of arterial and mixed venous blood respectively. As with the oxygen dissociation curve, this 'physio-logical' carbon dioxide dissociation curve moves between the two curves plotted with constant partial pressures of oxygen corresponding to oxygenated and deoxygenated blood.

globin for oxygen transport, is needed for carbon dioxide. For all partial pressures, the blood has a much greater capacity for carbon dioxide than for oxygen. This has certain important physiological consequences; for instance air which is introduced into a body cavity, or a bubble of gas forming within the body fluids, is reabsorbed.

Just as an increase in the partial pressure of carbon dioxide causes the oxygen dissociation curve of the blood to shift to the right, an increase in oxygen partial pressure causes the carbon dioxide dissociation curve to shift to the right, as shown in Figure 6.4. The difference is due to the different amounts of carbon dioxide carried by haemoglobin; more carbon dioxide can be carried by deoxygenated blood than by oxygenated blood. This is known as the **Haldane effect**. As with oxygen dissociation, the 'physiological' carbon dioxide dissociation curve for blood as it flows through systemic capillaries is steeper than the curve at constant oxygen partial pressure (Figure 6.6). Again the effect is a tendency to reduce the change in the partial pressure of carbon dioxide which would otherwise occur; for carbon dioxide, this has the additional advantage of minimizing the variation in pH of blood taking up carbon dioxide, as the next section describes.

THE RESPIRATORY QUOTIENT

The **respiratory quotient** (RQ) is the volume of carbon dioxide produced divided by the volume of oxygen consumed in the tissues in a given period of time. The value of the RQ depends on the nature of the metabolic energy source. For carbohydrates, the RQ is 1, for fats it is 0.7 and for proteins it is 0.8. A healthy subject on an ordinary mixed diet of carbohydrate, fat and protein has an RQ of typically 0.85. The RQ is of significance in its influence on the difference in pH comparing arterial and venous blood, to be described next.

THE DIFFERENCE IN pH OF SYSTEMIC ARTERIAL AND VENOUS BLOOD

When blood flows through the systemic capillaries, oxygen needed by the metabolizing tissues is unloaded from the blood and carbon dioxide produced by the metabolizing tissues is added to the blood. Each of these processes has implications for blood pH. Oxyhaemoglobin is more acid than deoxygenated haemoglobin; as a result the unloading of oxygen from blood in the systemic capillaries results in the blood becoming more alkaline. Carbon dioxide taken up by the blood becomes hydrated with the release of hydrogen ions; this tends to make the blood more acid as it traverses a systemic capillary. The two effects combined result in a fall of pH from 7.4 to 7.37. These data apply if the

RQ is 1. If the body is metabolizing fat and the RQ is 0.7, less CO_2 is evolved than O_2 consumed. In this situation, the change in pH due to uptake of CO_2 in the systemic capillaries is entirely offset by the change in configuration of haemoglobin. This is an elegant physiological adaptation in which off-loading of oxygen and uptake of carbon dioxide interact to minimize changes in the internal environment. (Section A.2.)

ERYTHROCYTES SWELL AS THEY TRAVERSE SYSTEMIC CAPILLARIES

The carbon dioxide taken up by the blood as it traverses the systemic capillaries diffuses through the plasma to the intracellular fluid of the erythrocytes where, with the help of the catalytic action of carbonic anhydrase, the majority of the CO_2 is hydrated to form hydrogen ions and bicarbonate ions. Most of the hydrogen ions are removed by buffering. The bicarbonate ions are not removed. Although many of them move into the plasma, this is a one-to-one exchange with chloride ions (Chapter 5); the total number of osmotically active particles in the erythrocyte cytoplasm has increased whereas the number of osmotically active particles in the plasma is unaltered. The intracellular osmotic pressure becomes slightly higher than the osmotic pressure of the plasma; under this osmotic gradient water moves into the erythrocytes causing them to swell. Hence the volume of erythrocytes in systemic venous blood is slightly greater than that of systemic arterial blood. The erythrocytes shrink again as they traverse the pulmonary capillaries, where the reverse chemical changes take place (Davson and Eggleton, 1962, p. 91).

CARBON DIOXIDE AND RESPIRATORY ACIDOSIS

The body produces carbon dioxide continuously at the rate of about 200 ml per minute in an adult at rest. Consider a subject who rebreathes his own air, as a result of being trapped in a small unventilated space or of being suffocated by criminal assault. The carbon dioxide content of the inspired gas progressively rises at a rate depending on the volume of the unventilated space. In the situation of zero respiratory exchange, e.g. with suffocation due to a plastic bag over the head, the partial pressure of carbon dioxide in the arterial blood rises at around 7 mmHg per minute (Figure 6.7). More important than the rise in P_{CO_2} is the fall in P_{O_2}, also shown in Figure 6.7. Three minutes of anoxia is sufficient to do irreparable damage to regions of the brain, four minutes results in irreversible death of the whole of the cerebral cortex and eight minutes kills the whole organism. In view of these facts, not a second should be lost in commencing artificial ventilation in an asphyxiated subject.

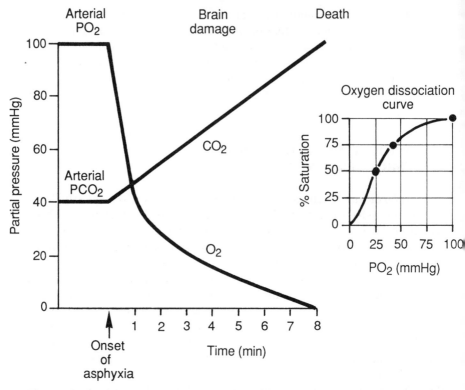

Figure 6.7. Partial pressures of oxygen and carbon dioxide in arterial blood (y-axis) as a function of time (x-axis). At time zero, ventilation ceases, e.g. as a result of a plastic bag put over the head. Inset: the oxygen dissociation curve shown for reference: see text for description.

The differences between the time course of fall in oxygen partial pressure and the rise in CO_2 partial pressure in asphyxia reflects the differences between the dissociation curves for the two gases. For CO_2, the dissociation curve is sufficiently near to being a straight line that, with CO_2 being released at a constant rate, the rate of rise of P_{CO_2} which is nearly linear. For oxygen, the dissociation curve is very non-linear. The following calculation demonstrates why the partial pressure of oxygen falls rapidly at first and then less rapidly as time progresses. Most of the oxygen stored in the body is in the blood in the form of oxyhaemoglobin. Typically, there is $5\frac{1}{2}$ litres of blood in a healthy adult; $5\frac{1}{2}$ litres of fully-oxygenated blood holds 1100 ml oxygen. In the body, only systemic arterial and pulmonary venous blood is fully oxygenated, so the amount of oxygen in the total blood volume is less than 1100 ml. At the start of asphyxia, in the unlikely event of the subject being in basal conditions, oxygen is used at typically 250 ml per minute; the blood holds enough oxygen for, at most, four minutes of oxygen consumption at this rate. In the first minute, the loss of a quarter of its oxygen causes a fall in Po_2 from 100 mmHg to 40 mmHg, a fall of 60 mmHg (see the oxygen dissociation curve inset in Figure 6.7). Removal of a further quarter of its oxygen will lower the Po_2 from 40 mmHg to 25 mmHg, a smaller fall of 15 mmHg. Other factors add to reducing the rate at which the Po_2 falls. As the Po_2 falls, cells reduce their usage of oxygen; the low Po_2 at the cell level results in aerobic metabolism giving way in some tissues to anaerobic metabolism. For these reasons, the rate of fall of Po_2 decreases as death approaches; it reaches zero after about eight minutes of asphyxia.

In hypoventilation, where aeration of the lungs is inadequate but not zero, the arterial partial pressure of carbon dioxide rises until the rate of excretion of carbon dioxide equals the rate of production (Figure 6.8). The ventilation is reduced but the concentration of carbon dioxide in the expired air rises. The

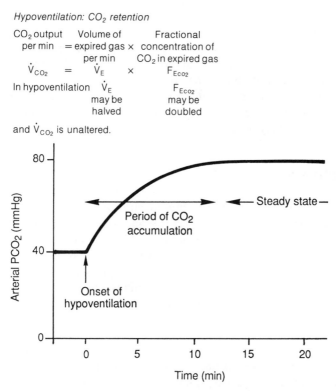

Figure 6.8. Partial pressure of carbon dioxide in arterial blood (y-axis) as a function of time (x-axis). At time zero, hypoventilation suddenly supervenes, e.g. as a result of a severe attack of asthma. As indicated, immediately following the onset of hypoventilation, there is a period during which carbon dioxide accumulates in the body. The Pco_2 rises exponentially towards a steady state value. When a steady state is established, the rate of carbon dioxide excretion in the lungs equals the rate of production in the tissues.

rate of carbon dioxide excretion is the product of expiratory minute volume and fractional pressure of carbon dioxide; the equation is shown in Figure 6.8. If the expiratory minute volume is halved, the excretion of carbon dioxide will be restored to normal if the fractional pressure of carbon dioxide in the expired air doubles. This doubling in partial pressure of carbon dioxide is reflected in the high partial pressure of carbon dioxide in the arterial blood which was the primary defect in respiratory acidosis.

SECTION 6.2 THE RESPIRATORY SYSTEM IN ACID–BASE PHYSIOLOGY

THE COMPOSITION OF ATMOSPHERIC AIR

Atmospheric air contains nitrogen, oxygen, carbon dioxide, trace concentrations of noble gases, such as argon, and water vapour. To a close approximation, atmospheric air consists of 80% nitrogen and 20% oxygen. The amount of carbon dioxide, about 0.04%, is small enough to ignore in all our considerations of mammalian respiration and metabolism. The amount of water vapour is much more variable than the concentration of any of the other components; it depends on atmospheric conditions but in a temperate climate it is quite small and can usually be ignored too.

INSPIRED AIR AND ALVEOLAR GAS

Atmospheric air is breathed in during inspiration. This 'inspired air' mixes with the gas already in the respiratory tree and some is drawn into the alveoli where gas comes into intimate relation with the blood circulating in the pulmonary capillaries. The gas here is called **alveolar gas**. Blood entering the lungs from the pulmonary artery has a partial pressure of oxygen which is lower than that in the alveolar gas. Oxygen diffuses from the alveolar gas into the pulmonary blood. For carbon dioxide, the partial pressure gradient is from blood to alveolar gas and so this is the direction in which carbon dioxide diffuses.

At the end of inspiration, the conducting airways (trachea, bronchi etc.) are full of gas similar in composition to air, except that the gas is saturated with water vapour. Since this gas does not take part in gaseous exchange with the blood, it is called the **dead space**. During expiration, gas is forced out of the lungs. The first gas to leave via nose and mouth is this dead space gas and it is followed by alveolar gas. This gas mixture can be collected and is known as **mixed expired gas**. Its composition is between that of air and alveolar gas, since it is a mixture of the two.

Alveolar gas contains the same gases, oxygen, nitrogen and carbon dioxide, as the inspired air but at different partial pressures. The partial pressure of oxygen is lower and that of carbon dioxide is higher; this is principally because, in the lung, oxygen is continually being removed from the alveolar gas by the blood and carbon dioxide is being added to the alveolar gas by the blood. The alveolar gas is saturated with water at body temperature. In any gas saturated with water vapour, the water vapour pressure depends only on the temperature. In the alveoli the temperature is held at 37°C by the body's temperature regulating mechanisms. At this temperature, the water vapour pressure is 47 mmHg.

The total pressure of gases in alveolar gas is equal to the barometric pressure. Since 47 mmHg is contributed by water vapour, the sum of the partial pressures of all other gases equals barometric pressure minus 47 mmHg.

THE PARTIAL PRESSURE OF GASES IN ALVEOLAR GAS AND IN SYSTEMIC ARTERIAL BLOOD

In health, the alveolar membrane which separates the blood in the pulmonary capillaries and the alveolar gas is so thin that the blood equilibrates with alveolar gas in its transit through the lungs. The partial pressures in systemic arterial blood are within a few mmHg of those in alveolar gas. In the context of acid–base physiology, the partial pressures may be taken as being equal, although there are situations, such as thickening of the alveolar membrane or ventilation-perfusion imbalance, where this near equality is not maintained (Jennett, 1989, p. 217).

THE BODY REGULATES THE PARTIAL PRESSURE OF CARBON DIOXIDE IN THE SYSTEMIC BLOOD

In a healthy person breathing air or any other gas mixture containing adequate oxygen, powerful homeostatic mechanisms regulate respiration in such a way that the partial pressure of carbon dioxide in the systemic arterial blood is stabilized close to 40 mmHg. If the metabolic production of carbon dioxide falls or rises, alveolar ventilation is changed in exact proportion so that the partial pressure of carbon dioxide is held constant. For instance, if a healthy person breathing air exercises vigorously, the metabolic production of carbon dioxide may rise from a resting value of 250 ml per min. to ten times that value. Accompanying this, alveolar ventilation is also increased ten-fold and the alveolar partial pressure of carbon dioxide remains close to 40 mmHg.

SO LONG AS HYPOXIA IS EXCLUDED, THE BODY DOES NOT REGULATE THE PARTIAL PRESSURE OF OXYGEN IN THE SYSTEMIC BLOOD

So long as there is adequate oxygen for metabolic needs, the partial pressure of oxygen in the systemic arterial blood is not regulated. A normal person at rest breathing air at a barometric pressure of one atmosphere pressure has a partial pressure of oxygen in his alveolar gas of typically 100 mmHg. If the subject breathes pure oxygen at the same barometric pressure, the partial pressure of oxygen in the systemic arterial blood rises to above 600 mmHg whilst the partial pressure of carbon dioxide in arterial blood remains at 40 mmHg.

THE PARTIAL PRESSURE OF NITROGEN

Nitrogen is neither absorbed nor excreted in the lungs. Consequently, the total amount of nitrogen inspired and expired per minute is the same. However, the partial pressure of nitrogen is different comparing air with alveolar gas because of the evaporation of water from the respiratory tract into the inspired air. This adds water vapour to the alveolar gas until a partial pressure of 47 mmHg is achieved, with the effect of lowering the partial pressure of nitrogen.

In order to understand this, let us consider an initially dry gas at one atmosphere pressure in an enclosed chamber fitted with a plunger. The movement of the plunger maintains the total pressure at atmospheric. Suppose the chamber and contents are at 37°C and that a drop of water is introduced into the container. As water evaporates to saturate the gas, the total volume of the gas must be increased in order to prevent the total pressure from rising. Thus the partial pressures of all other gases in the container are reduced in proportion to the increase in volume, even though these gases are being neither removed from nor added to the gas mixture. Similarly, for a person in a steady state, the partial pressure of nitrogen in the alveolar gas is less than that in inspired air in the proportion (barometric pressure : barometric pressure minus 47 mmHg).

THE SEE-SAW OF CARBON DIOXIDE AND OXYGEN

This is a useful concept which applies to subjects breathing air or any other gas mixture free of carbon dioxide. It relates the partial pressure of oxygen in alveolar gas to that of carbon dioxide. Oxygen and carbon dioxide are continually moving between the alveolar gas and the blood in the pulmonary capillaries.

The steps leading to the see-saw concept are as follows:

1. For simplicity, the case is considered where the amount of oxygen moving from the alveolar gas to the blood equals the amount of carbon dioxide moving to the alveolar gas from the blood. When these amounts of oxygen leaving alveolar gas and of carbon dioxide entering it are not equal, the pivot of the see-saw, to be described, is moved to the left to incorporate this complicating factor.
2. In any gas mixture, the partial pressure exerted by one component of the gas is directly proportional to the concentration of that component.
3. Steps 1 and 2 together show that, in the alveolar gas, the rise in partial pressure of carbon dioxide resulting from CO_2 entering is the same as the fall in partial pressure of oxygen resulting from oxygen leaving.
4. In alveolar gas, the P_{O_2} is lower than in inspired air by the same amount as the P_{CO_2} is higher.

The see-saw diagram (Figure 6.9A) is a graphic representation of this. The diagram indicates the composition of alveolar gas of a person at sea level breathing air. Carbon dioxide is shown on the left and oxygen on the right. The vertical axis is calibrated both in mmHg and as percentage concentration, assuming a total atmospheric pressure of one atmosphere. The ends of an oblique line passing through the pivot point indicate corresponding values for carbon dioxide on the left and oxygen on the right. See-saw movements of this line correspond to changes in ventilation. For a normal subject breathing air, the corresponding see-saw line is labelled 'eupnoeic', meaning 'good respiration', the respiration being appropriate to the metabolic rate. The concentration of carbon dioxide is 5.2% (partial pressure 40 mmHg) and of oxygen is 13.2% (partial pressure 100 mmHg). It is easier to work in partial pressures than in percentages because the partial pressures are round numbers. For the line representing a hyperventilating subject, the P_{CO_2} is 20 mmHg and the P_{O_2} is 120 mmHg. For the line representing the hypoventilating patient, the $P_{CO_2} = 80$ mmHg and the $P_{O_2} = 60$ mmHg. The line labelled 'air' shows the composition of air at one atmosphere when saturated with water vapour at a temperature of 37°C, the P_{CO_2} being too small to distinguish from zero and the P_{O_2} being 140 mmHg.

IN HYPERVENTILATION, THE BODY IS DEPLETED OF CARBON DIOXIDE BUT IS NOT LOADED WITH EXTRA OXYGEN

Figure 6.9B, which shows the oxygen and carbon dioxide dissociation curves for blood together, illustrates how much more carbon dioxide than oxygen is held by the blood. The points corresponding to the eupnoeic state are labelled 'Normal'; those for hypoventilation are labelled 'Hypo' and those for hyper-

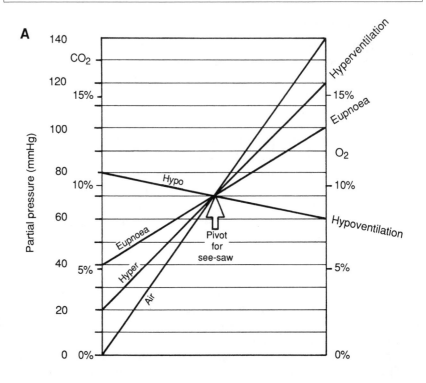

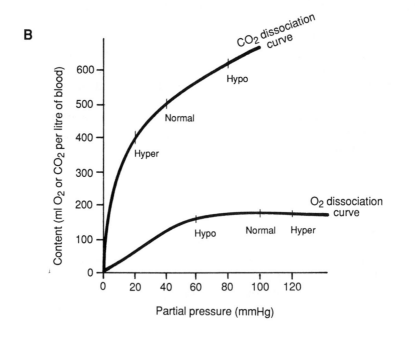

ventilation are labelled 'Hyper'. Comparing hyperventilation with normal, the haemoglobin is saturated with oxygen in both states; changes in partial pressure of oxygen have insignificant effects on blood content of oxygen. For the carriage of carbon dioxide, by contrast, no such saturation phenomenon exists; changes in partial pressure are accompanied by relatively large changes in blood content.

The state represented by the points 'Hyper' in Figure 6.9A and B is applicable only when the overbreathing has been established for long enough for a new steady state to have become established. At the onset of hyperventilation, the systemic arterial oxygen partial pressure rises within seconds to its final value. The partial pressure of carbon dioxide falls more slowly as carbon dioxide is washed out of the body. During this phase, the body is excreting more carbon dioxide than it is producing. This phenomenon leads to a new definition: the **respiratory exchange ratio** is the CO_2 output through the lungs divided by the oxygen uptake through the lungs. At times when the excretion of CO_2 matches its production, the respiratory exchange ration equals the RQ. If excretion exceeds production, the respiratory exchange ratio exceeds the RQ. In the early phases of hyperventilation the respiratory exchange ratio may approach 2. Over the course of approximately a quarter of an hour the carbon dioxide partial pressure falls asymptotically to its new value and a new steady state is reached where carbon dioxide excretion once again matches its metabolic production. The same applies in reverse to rapidly developing hypoventilation.

BLOOD OXYGENATION AND DISORDERS OF ACID–BASE PHYSIOLOGY

In considering hypoventilation earlier in this book, attention was focussed on the rise in carbon dioxide partial pressure since this is the factor of importance in acid–base considerations. In the present context, it is clear that a rise in the blood carbon dioxide partial pressure is necessarily accompanied by a fall in oxygen partial pressure. Of the two, the fall in oxygen partial pressure is much more damaging to the organism because cells cannot survive without oxygen.

There are respiratory disorders in which the blood oxygen partial pressure falls with insignificant change or even a fall in the partial pressure of carbon

---

Figure 6.9. A. The 'see-saw' of partial pressures of carbon dioxide and oxygen in respiratory gases. This is a diagrammatic representation of the fact that, comparing the partial pressures of gases in inspired air with alveolar gas, the P_{O_2} decreases by the same amount as the P_{CO_2} rises. B. Carbon dioxide and oxygen dissociation curves plotted together. This emphasizes a) the greater capacity of blood for carbon dioxide than for oxygen, and b) as the P_{O_2} of arterial blood increases above the normal value of 100 mmHg, almost no extra oxygen is taken up whereas for CO_2 there is no such saturating effect.

dioxide. This occurs when diffusion across the alveolar wall is impaired as a result, for instance, of thickening of the wall. Since oxygen diffuses much less readily than does carbon dioxide, the diffusion barrier causes a large difference in oxygen partial pressure between alveolar gas and pulmonary capillary blood with almost no effect on the carbon dioxide differential. The hypoxic stimulus to ventilation, if sufficiently severe, causes hyperventilation so that a low P_{CO_2} accompanies the low P_{O_2}. A progressive thickening of the alveolar wall kills the patient from arterial hypoxia long before diffusion of carbon dioxide is compromised.

In a normal person at high altitude, where the atmospheric pressure is reduced, hypoxia provides a drive to respiration. This leads to hyperventilation. In this condition, both the alveolar P_{O_2} and P_{CO_2} are lower than in the subject at sea level.

THE CONTROL OF VENTILATION

Ventilation is produced by the muscles of respiration which are in turn controlled by the respiratory neurones most of which lie deep in the medulla oblongata. These are stimulated by 'central chemoreceptors' which lie close to the ventral surface of the medulla. In a person breathing air, the respiratory system controls the partial pressure of carbon dioxide in the systemic arterial blood. The principal mechanism is via the hydrogen ion concentration of the cerebral interstitial fluid bathing the central chemoreceptors. Carbon dioxide, being an uncharged molecule, dissolves readily in lipids; it crosses the blood-brain barrier with ease by diffusing through the lipid membrane of the endothelial cells of the cerebral capillaries. Ions such as bicarbonate and hydrogen ions, being polar, are lipid-insoluble; they are not able readily to cross the blood-brain barrier. When the partial pressure of carbon dioxide in the blood perfusing the medulla varies from its usual value, this results in an immediate change in the partial pressure of carbon dioxide in the cerebral interstitial fluid. This fluid is poorly buffered, being almost devoid of protein, and consequently, for a given change in carbon dioxide partial pressure, the interstitial fluid exhibits a much larger change than plasma in hydrogen ion concentration. The central chemoreceptors are specifically sensitive to changes in the hydrogen ion concentration of their environment. Acidity stimulates respiration and this is the origin of the increase in respiratory drive when the arterial carbon dioxide partial pressure rises.

A lowering of the oxygen partial pressure in arterial blood is also a respiratory stimulant. The receptors are the peripheral chemoreceptors; they are highly vascular structures lying close to the bifurcation of the common carotid arteries (the carotid bodies) and close to the aortic arch (aortic bodies). When the blood is in its normal state of oxygenation, these receptors show a low rate of firing. Any reduction of the oxygen partial pressure below its usual value of 100 mmHg stimulates these receptors to fire more frequently, but it requires a fall in oxygen partial pressure to around 60 mmHg before this stimulus predominates over the maintenance of a normal Pco_2.

Whereas the central chemoreceptors are little influenced by the hydrogen ion concentration of the blood itself, because hydrogen ions do not readily cross the blood-brain barrier, the peripheral chemoreceptors are sensitive to the hydrogen ion concentration and the Pco_2 of the arterial blood in addition to their sensitivity to hypoxia. Although a fall of Po_2 alone must be to 60 mmHg before there is much of a reflex increase in ventilation, when a fall in Po_2 is accompanied by a rise in Pco_2 (as in asphyxia or breath-holding), these receptors respond by firing very frequently and a large stimulation of ventilation is produced.

In metabolic acidotic disorders of acid–base physiology, it is the peripheral chemoreceptors which sense the pH of the extracellular fluid and cause the hyperventilatory compensatory response.

APPENDIX 6

A.1 'Oxygen dissociation' curve of the blood

The amount of oxygen which a litre of blood yields as a function of the partial pressure of oxygen with which it is equilibrated is called the 'oxygen dissociation curve of the blood' in this book and by many authorities. Its meaning is clear; it is the total amount of oxygen held in blood, be it in chemical combination with haemoglobin or in physical solution. Some authors claim that this is a misnomer since the oxygen does not itself dissociate; most of the oxygen is derived from the dissociation of oxyhaemoglobin. This leads to other expressions such as the 'oxyhaemoglobin dissociation curve' or the oxygen-haemoglobin dissociation curve. This nomenclature carries its own scientific penalty, since, strictly applied, it refers only to oxygen associated with haemoglobin and excludes oxygen in physical solution.

A.2 The pH of blood as it passes through a systemic capillary

Haemoglobin is an effective buffer in the pH range 7.0 to 7.6. Changes in hydrogen ion concentration, resulting from changes in the partial pressure of carbon dioxide as blood flows through capillaries, are largely buffered by ionizing groups of haemoglobin which are influenced by the oxygenation and deoxygenation of the haem groups. On oxygenation, these groups release hydrogen ions; the oxyhaemoglobin is thus a stronger acid than haemoglobin. When oxyhaemoglobin dissociates to haemoglobin and oxygen, hydrogen ions are taken up again by the haemoglobin. To keep the pH constant when one gram-molecule of oxygen dissociates from oxyhaemoglobin, 0.7 gram-molecules of hydrogen ions must be added. This amount of hydrogen ions would be provided if 0.7 gram-molecules of CO_2 were completely hydrated.

When the erythrocyte traverses a systemic capillary in a tissue with an RQ of 0.7, the loss of oxygen by the oxyhaemoglobin is exactly offset by the uptake of hydrogen ions from the hydration of carbon dioxide; the internal pH of the erythrocyte is unchanged. At higher RQs the remainder of the CO_2 releases hydrogen ions which are buffered by the blood buffers. In the lungs, the reverse changes take place.

From the passive buffer properties of the blood, with an RQ of one, the predicted change in pH of the blood as a result only of the uptake of CO_2 would be from 7.4 to 7.32. However, as a consequence of the alkalinizing effect of dissociation of oxyhaemoglobin to oxygen and haemoglobin, the actual fall in pH is only to 7.37.

Renal aspects of acid–base physiology

The function of the kidney is to maintain the constancy of the internal environment and it contributes to correction of changes in acid–base status of the blood by regulating the excretion of acid or alkali. The limits of pH of urine are far wider than the limits for plasma; they are 4.5 pH units on the acid side and 7.8 pH units on the alkaline side.

THE MECHANISM OF THE RENAL SECRETION OF HYDROGEN IONS

In a human on a mixed diet, the urine is acid. The glomerular filtrate has the same hydrogen ion concentration as the plasma and the acidification of the urine is accomplished by renal tubular pumping of hydrogen ions into the tubular fluid. The mechanism is rather complicated, so it is shown in two stages in Figure 7.1. Figure 7.1A concerns the production of hydrogen ions in the tubular cell; Figure 7.1B shows how electroneutrality is maintained.

Urinary hydrogen ions are derived from carbon dioxide produced by metabolism (Figure 7.1A). The CO_2 reacts with water to form hydrogen ions and bicarbonate ions as shown by the reaction inside the cell. This reaction, as previously noted, occurs very slowly unless the enzyme carbonic anhydrase is present to catalyse it. The renal tubular cells are rich in this enzyme. The importance of carbonic anhydrase was established in a series of classical studies by Pitts (1948, 1974); one line of evidence was the observation that inhibition of carbonic anhydrase abolished renal excretion of acid.

Hydrogen ions produced by the reaction are transferred from the tubular cell cytoplasm into the tubular fluid. The mechanism involves a metabolically-driven pump. This removal of hydrogen ions from the intracellular fluid causes the chemical reaction to move to the right by a mass action effect, thus replenishing the supply of hydrogen ions inside the cell. This process utilizes intracellular

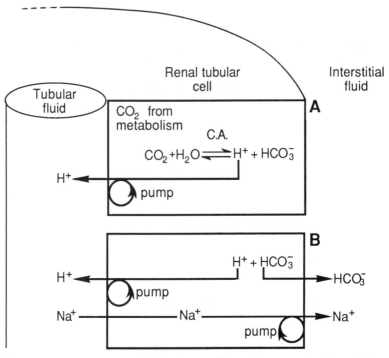

Figure 7.1. The renal tubular mechanism for the excretion of acid. For clarity, the mechanism is shown in two stages in parts A and B. A. CO_2 from metabolism is hydrated; the released hydrogen ions are pumped into the tubular fluid. B. The accompanying transmembranal movements of sodium and bicarbonate are shown.

carbon dioxide which is supplied by general metabolic processes within the tubular cells, by cells elsewhere in the body and from tubular bicarbonate, by means of a mechanism to be described shortly. Thus hydrogen ions secreted into the tubular fluid are derived from carbon dioxide.

For each hydrogen ion excreted, one bicarbonate ion is added to the intracellular fluid of the renal tubular cell. The intracellular concentration of bicarbonate rises. In Figure 7.1B, the reaction of CO_2 is omitted and only the products H^+ and HCO_3^- are reproduced from Figure 7.1A. Bicarbonate, being charged, is insoluble in lipid and so diffuses extremely slowly across the lipid regions of the cell membrane. On the aspect of the tubular cell membrane facing the interstitial fluid, the lipid membrane is traversed by protein macromolecules which comprise a carrier mechanism for bicarbonate ions (section A.1). The rise in intracellular concentration of bicarbonate leads to transfer of bicarbonate from the renal tubular cell cytoplasm to the renal interstitial fluid.

Sodium movements accompany these movements of hydrogen and bicarbonate ions. In the proximal convoluted tubule, on the tubular fluid aspect of the cell, sodium ions move from urine into the cell. On the aspect of the tubular cell facing the interstitial fluid, sodium is pumped actively out of the cell. Reference to Figure 7.1B allows a rough check on electroneutrality. To the left of the tubule cell, sodium and hydrogen ions, both positively charged, exchange; on the right aspect, sodium ions and bicarbonate ions, oppositely charged, both leave the cell.

Bicarbonate is freely filtered in the glomeruli and the mechanism for its retrieval is intimately connected with hydrogen ion secretion already described. Again the mechanism is described in two steps, shown in Figures 7.2A and B. Figure 7.2A shows the essential components of the mechanism. Bicarbonate is freely filtered at the glomerulus, its concentration in glomerular filtrate being equal to that in plasma. In the tubular fluid, the secreted hydrogen ions combine with the bicarbonate ions to form carbon dioxide and water. The partial pressure of carbon dioxide in the tubular fluid rises and CO_2 diffuses into the tubular cell. This CO_2 then feeds back into the mechanism for generating hydrogen ions, as shown in the lowest chemical reaction of Figure 7.2B. This reaction is none other than the top reaction of Figure 7.1A; a complete cycle has been achieved.

In glomerular filtration, the charge of each filtered bicarbonate ion is balanced by that of a sodium ion; each bicarbonate ion in the filtrate can be regarded as being accompanied by a sodium ion, although the two are not chemically bonded. The sodium was omitted from Figure 7.2A for clarity but it is shown in Figure 7.2B. The bicarbonate ions are retrieved by the CO_2 diffusion mechanism just described and the sodium ions are reabsorbed. The net effect of all these steps is the equivalent of net transfer of sodium ions and bicarbonate ions from filtered fluid back into the renal venous blood, i.e. renal reabsorption of the filtered sodium bicarbonate. During reabsorption, sodium ions retain their identity but bicarbonate ions do not. At no stage do bicarbonate ions themselves cross the aspect of the renal tubular cell membrane facing the tubular fluid; the bicarbonate ions which are filtered are not the same as the ones which enter the renal venous blood. For this reason, some authorities prefer to use the expression 'renal retrieval of bicarbonate' rather than 'renal reabsorption of bicarbonate'. The transfer of the bicarbonate is indicated by the tube in Figure 7.2B.

Figure 7.3 illustrates the fact that, as the tubular fluid proceeds along the nephron, bicarbonate is progressively retrieved until the tubular fluid is virtually bicarbonate-free when, as is usually the case, the urine is acid. Around 90% of filtered bicarbonate is retrieved in the proximal convoluted tubule. Along this first length of the tubule, the hydrogen ions secreted into the tubular fluid as a step in this mechanism accomplish the task of retrieving filtered bicarbonate. Hydrogen ion secretion here does not result in the excretion of acid. It is only as the tubular fluid proceeds further that renal tubular secretion of hydrogen ions does result in acid excretion.

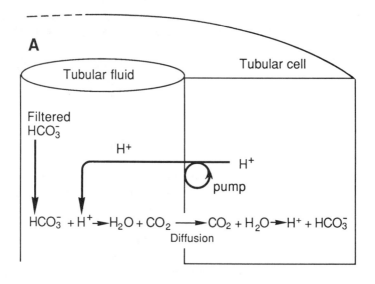

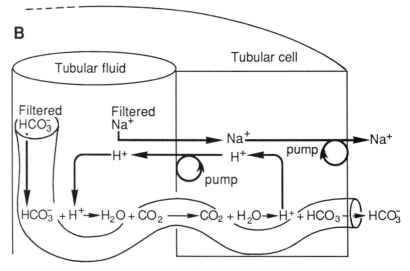

Figure 7.2. The renal tubular mechanism for the retrieval of bicarbonate. For clarity, the mechanism is shown in two stages in parts A and B. Bicarbonate filtered in the glomeruli reacts with hydrogen ions secreted into the tubular fluid to yield CO_2 which diffuses into the tubular cell cytoplasm to react with water. B. The accompanying transmembranal movements of sodium and bicarbonate are shown. The pathway taken by bicarbonate ions is indicated by the tube, to help in emphasizing the retrieval of filtered bicarbonate ions. The part of the figure representing sodium reabsorption is more obvious to the eye and so has not been provided with a tube. Hydrogen ions in this first segment of the tubule move in a closed loop as an essential part of the mechanism for retrieval of bicarbonate. As with bicarbonate, they lose their identity during the process. For hydrogen, this loss of identity is at the stage of movement from tubular fluid to tubular cell cytoplasm. The hydrogen atom has become incorporated in a water molecule, which moves from tubular fluid to tubular cell cytoplasm; this movement is not shown in the figure.

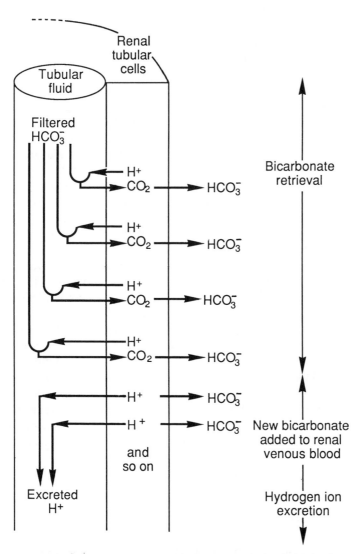

Figure 7.3. Tubular hydrogen ion secretion into the tubular fluid: in the proximal length of the tubule, hydrogen ion secretion is used up in retrieval of bicarbonate whereas, in the distal length of the tubule, the hydrogen ions are excreted. In the proximal tubule, the bicarbonate entering the interstitial fluid is retrieved from the tubular fluid whereas, in the distal tubule, it is 'new' bicarbonate derived from CO_2.

HYDROGEN ION SECRETION ALONG THE NEPHRON

Hydrogen ion secretion into the tubular fluid occurs along the whole length of the proximal and distal convoluted tubules and in the collecting ducts. As the tubular fluid proceeds along the nephron, the concentration of bicarbonate falls and a progressively greater proportion of the secreted hydrogen ions remains in the tubular fluid, to be buffered by phosphate, ammonia etc. In the proximal tubule, the pH of the tubular fluid is only slightly less than that of plasma and a large quantity of H^+ is pumped against a low gradient. In the distal tubule, the pH of the tubular fluid is low and a small quantity of H^+ is pumped against a large gradient, this pumping requiring energy derived from metabolism. The pH of the tubular fluid progressively falls both because of addition of hydrogen ions H^+ to the tubular fluid and because of the removal (reabsorption) of water from the tubular fluid. The number of hyrogen ions secreted depends on the acid–base status of the subject and on the health of the kidneys.

In the proximal tubule, the bicarbonate entering the renal interstitial fluid and thence being removed by the renal venous blood arises indirectly from filtered bicarbonate. In the distal lengths of the tubule, the hydrogen ions secreted into the urine stay there whilst the bicarbonate ions added from the renal tubular cell cytoplasm to the interstitial fluid are now 'new' bicarbonate ions, not merely ions retrieved from the tubular fluid.

Sodium ion movements have been omitted from Figure 7.3 for clarity; at every step where bicarbonate is shown moving from the tubular cell cytoplasm to the renal interstitial fluid, there is an accompanying sodium ion derived from renal tubular fluid. Other ions contributing smaller amounts of charge have also been omitted. When the movements of all ions (e.g. Na^+, K^-, HCO_3^-) are added, there is an exact balance in charge both in the movements of ions between the tubular fluid and the tubular cell cytoplasm, and between the tubular cell cytoplasm and the renal interstitial fluid. Also in the urine itself, of course, there is always an exact balance of ionic charges. Here phosphate and ammonium ions are important in addition to sodium etc.

EXCRETION OF ACID

There is a limit to the magnitude of the gradient against which pumping can occur; this limit is reached when the pH of the tubular fluid falls to around 4.5. If all the hydrogen ions excreted from the body via the urine were freely ionized in the urine, the amount of acid which could be excreted would be limited to the amount present in solution with a pH of 4.5 and this is 30 micromoles of hydrogen ions per litre of urine. If in 24 hours $1\frac{1}{2}$ litres of urine is excreted, this corresponds to a total daily excretion of 45 micromoles of hydrogen ions. Metabolism in a person on a typical omnivorous diet yields typically 30 millimoles of acid from sulphur containing amino acids and this must be excreted in the urine if acid accumulation in the body is to be avoided. So the kidney must excrete the majority of the acid in a buffered state, i.e. associated with bases.

PHOSPHATE AS A BUFFER OF URINE

Phosphate makes a major contribution to the buffering of urine. Phosphates are derived from the products of metabolism and, on a typical western diet, 30 millimoles of phosphate, mainly dihydrogen phosphate, is excreted in the urine in 24 hours.

The buffer reaction is:

$$H_2PO_4^- = H^+ + HPO_4^{2-} \qquad \text{reaction 1}$$

The pK for this buffer pair is 6.8. We write the Henderson-Hasselbalch equation for this reaction as equation 1:

$$pH = 6.8 + \log([HPO_4^{2-}]/[H_2PO_4^-]) \qquad \text{equation 1}$$

When the phosphate enters the tubular fluid at a pH of 7.4, about three-quarters of the phosphate is in the form of monohydrogen phosphate, the base form. When the tubular cells are pumping hydrogen ions into the luminal fluid, the pH falls and reaction 1 is driven, by mass action effects, to the left. At the lowest possible tubular fluid pH of 4.5, almost all the phosphate is in the form of the dihydrogen phosphate, the acid form. So the phosphate has been almost quantitatively changed from base to acid. For every mole of phosphate entering the tubular fluid, about three-quarters of a mole of hydrogen ions is buffered. In more normal circumstances when the urine is not so extremely acid, phosphate acts as a buffer for a relatively smaller amount of acid; the exact amount can be readily calculated from the Henderson-Hasselbalch equation and the pH of the urine.

Since a typical yield per day by metabolism of both phosphate and acid is 30 millimoles, phosphate is present in sufficient amount to carry away all the acid. In normal circumstances, all of its buffering power is not used (the pH of urine is not at the lowest possible value) and some acid is excreted carried by ammonia. When extra acid is released into the body by an unduly large intake of acid by mouth or by pathological conditions, the phosphate system alone would be inadequate and carriage by ammonia is mandatory if the body is to rid itself of this excess acid.

AMMONIA

Ammonium (NH_4^+) is a very weak acid of the strong base ammonia (NH_3). The pK is 9.4 (see Figure 1.4A); consequently the ammonium/ammonia pair cannot function as an important buffer in the conventional sense in the body fluids. It plays a specialized role as a major vehicle in carrying hydrogen ions in the urine and hence excreting acid from the body, this role being dependent on the high permeability of lipid membranes to ammonia but not ammonium.

Ammonia formation in the body

Ammonia is produced in proximal and distal segments of the tubules. It is formed by the deamination of glutamine and other amino acid substrates in the liver, intestinal mucosa and the kidney. Any ammonia in the blood is taken up by the liver and converted to urea, the liver being the only organ in which urea is formed. Unlike blood, the urine contains appreciable quantities of ammonia. The enzyme **glutaminase**, located in the mitochondria of the renal tubule cells, catalyses the production of ammonia; the reaction is shown in section A.2. Glutaminase is present in large amounts in the kidney and its concentration there is raised in acidosis.

Ammonia and renal acid–base physiology

Ammonia is an exceptional chemical in that it is manufactured in the renal tubular cells and excreted into the urine. For most other chemicals, the kidney operates by transferring them in one direction or the other between the tubular and peritubular fluids but without itself manufacturing the chemicals. The ammonia produced in the tubular cells is derived from glutamine, as already described. The other chemicals yielded by the splitting of ammonia from glutamine are neither acids nor alkalis; this mechanism provides the kidney with a source of base which can be produced without incurring other acid–base alterations.

In renal physiology, it is conventional to divide the renal mechanisms for the production of urine into three groups, glomerular filtration, tubular secretion and tubular reabsorption. The unique status of ammonia requires a fourth mechanism in renal physiology, which is the addition of newly-produced chemical to the urine.

The role of ammonia is in buffering acid and it makes a contribution which is particularly important when the acid load is high. Ammonia produced in the tubular cell cytoplasm diffuses readily across the tubular cell membrane into the tubular fluid. Cell membranes consist largely of lipid material and so are in general permeable to non-polar particles such as ammonia The solubility of ionized particles such as ammonium ions in lipid is very low so that ions permeate cell membranes very slowly.

The ammonia, once it diffuses into the luminal fluid, reacts with hydrogen ions to produce ammonium ions, according to the reaction

$$NH_3 + H^+ = NH_4^+$$

When the tubular cells are vigorously transporting hydrogen ions into the tubular fluid as a compensation for acidosis, the concentration of hydrogen ions in the tubular fluid rises. This, by a mass action effect, drives the reaction to the right. This association of hydrogen ions with ammonia removes the hydrogen ions from the ionized pool and is thus an effective vehicle for conveying hydrogen ions in the urine without undue lowering of the urinary pH.

Because of the different permeabilities of the tubular cell membrane to ammonia molecules and ammonium ions, the ammonium ions are trapped in the tubular fluid. When the tubular fluid is acid, the move of the reaction to the right exaggerates the trapping effect. Consequently the rate of ammonium excretion in the urine increases with the hydrogen ion concentration of the urine. This relationship is shown in Figure 7.4. For the normal subject, as the pH falls, there is a rise in the rate of excretion of ammonium and the ammonium carries much of the acid load. In a normal healthy person about twice as much acid can be excreted at a urinary pH of 4.5 than would be the case if the ammonia mechanism were not present.

The stimulus to production of ammonia is intracellular acidosis in renal tubular cells. When an acidosis develops and persists, the rate of ammonia production by the renal tubular cells increases over the course of several days. This is due to induction of the enzyme glutaminase (i.e. the production of new enzyme). The rate of production of ammonia provides a homeostatic mechanism in the excretion of excess acid from the body. As is shown in Figure 7.4, in chronic metabolic acidosis the rate of ammonium ion excretion in the urine at a given urine pH is more than double the rate in a healthy person (Pitts, 1948).

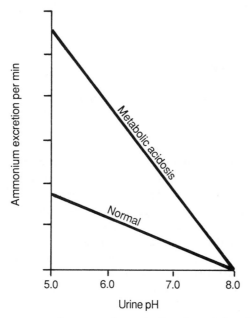

Figure 7.4. Urinary ammonium excretion as a function of the pH of the urine in normal subject and in chronic acidosis.

TITRATABLE ACID AND TOTAL ACID EXCRETION

Titratable acid: the 'titratable acid' of the urine is measured by titrating the urine produced in a given period of time back to the pH of the blood. This gives the number of hydrogen ions free in the urine and buffered by the urinary buffers such as phosphate.

Since ammonia is added to the urine by the renal tubular cells themselves, the amount of urinary acid buffered as ammonium ions is not included in the titratable acid. To estimate the contribution of ammonia, we must directly measure the amount of ammonium ions in the urine and add this to the titratable acid. This is called the **total acid excretion**:

RENAL FAILURE

In this condition, all metabolic functions of the kidney are depressed; these functions include tubular secretion of hydrogen ions, reabsorption of bicarbonate ions and production of ammonia. In a subject with impaired renal function, the urine can scarcely be concentrated or diluted by comparison with plasma, its pH can be only slightly lowered below or raised above the pH of plasma and, because of lack of ammonia synthesis, the excretion of acid is profoundly depressed. The kidney can no longer perform its homeostatic regulatory role. Such a patient consuming a normal diet becomes progressively more acidotic because of the release of acid resulting from the metabolism of protein (Chapter 5).

DIURETIC THERAPY

Diuretic therapy is a common iatrogenic origin of metabolic disturbances of acid–base physiology. Diuretics are administered for their naturetic properties particularly in patients with cardiac, hepatic, pulmonary and renal disease, to rid the body of excess extracellular fluid. When the loss of sodium is matched by losses of other extracellular electrolytes in proportion to their extracellular concentrations, no disturbance of acid–base balance occurs. In cases where there is a disproportionate loss of bicarbonate, the result is metabolic acidosis. Conversely when there is an exaggeration of loss of ammonium or chloride ions by comparison with sodium, this leads to metabolic alkalosis.

A contribution to the mechanism of increased acid excretion with diuretics is that the diuretic induces an increase in aldosterone secretion (Cohen and Kassirer, 1982, p. 261). This promotes sodium reabsorption in the renal tubules and an accompanying aciduria, as described in Chapter 3.

Carbonic anhydrase inhibitors are used in the treatment of glaucoma and, less commonly, in the treatment of certain types of stones in the renal tract.

This treatment causes a mild metabolic acidosis, due to impairment of renal reabsorption of bicarbonate. These agents are only moderately potent diuretics. Their diuretic effect is due to this impairment of bicarbonate reabsorption and as metabolic acidosis develops, the plasma bicarbonate concentration falls. The filtered load of bicarbonate therefore falls also and the diuresis is therefore poorly maintained.

APPENDIX 7

A.1 The bicarbonate–sodium co-transporter for movement from the renal tubular intracellular fluid to the renal interstitial fluid

This specific carrier is of such a configuration that it will only allow bicarbonate ions to cross the membrane if accompanied by sodium ions. It is a passive carrier, i.e. it does not require energy. The source of energy is the increased outward concentration for bicarbonate arising from intracellular hydration of carbon dioxide, as described in the text. Two or three bicarbonate ions are transported for each sodium ion.and it is because of this that the transport mechanism can operate without a metabolic pump being directly involved. The gradient driving each ion of bicarbonate outwards is considerably less than the gradient driving each sodium ion inwards; the loss of energy by one bicarbonate ion moving out would not be enough to provide the energy needed to carry a sodium ion outwards. By linking the movement of one sodium ion with that of two or three bicarbonate ions, the transfer mechanism becomes energetically feasible, with the summed energy loss associated with the transfer of several bicarbonate ions being sufficient to provide the driving force for a single sodium ion.

Sodium extrusion is also effected by the sodium pump, powered by energy from metabolism, which pumps sodium ions out of the cell into the interstitial fluid.

A.2 The production of ammonia from glutamine

$$
\begin{array}{ccccccc}
CONH_2 & & & & COOH & & \\
| & & & & | & & \\
CH_2 & & & & CH_2 & & \\
| & & & & | & & \\
CH_2 & + & H_2O & \underset{\text{Glutaminase}}{\rightleftharpoons} & CH_2 & + & NH_3 \\
| & & & & | & & \\
CHNH_2 & & & & CHNH_2 & & \\
| & & & & | & & \\
COOH & & & & COOH & & \\
\textbf{Glutamine} & \textbf{Water} & & & \textbf{Glutamic} & \textbf{Ammonia} \\
& & & & \textbf{acid} & &
\end{array}
$$

Self-test questions and answers 8

QUESTIONS

CHAPTERS 1 AND 2

Structured exercise 1 Acid–base status – introduction to its graphic representation

Review of material from Chapters 1 and 2.

Write down the chemical reaction(s) occurring when there is a change in the P_{CO_2} with which an aqueous solution is equilibrated.

From the law of mass action, show that:

$$[H^+] = K[CO_2]/[HCO_3^-] \quad \text{(equation 1)}$$

$[CO_2]$ represents the concentration of CO_2 in physical solution.

When $[CO_2]$ and $[HCO_3^-]$ are measured in M, K is determined experimentally as being 0.8×10^{-6} M.

Why is K given the dimensions of concentration?

A. Graphic representation
We can choose any two of the three variables $[H^+]$, $[CO_2]$, $[HCO_3^-]$ to represent acid–base status. On the outline graph (Figure 8.1), $[HCO_3^-]$ will be plotted as a function of P_{CO_2} or $[CO_2]$.

Look at the axes of the graph. Why is it possible to label the x-axis either in units of P_{CO_2} (mmHg) or in CO_2 concentration $[CO_2]$?

At any point on the graph, $[H^+]$ is determined by equation 1.

During this exercise, you will be invited to consider particular points on the graph of Figure 8.1. These points have been chosen to keep the calculations as simple as possible.

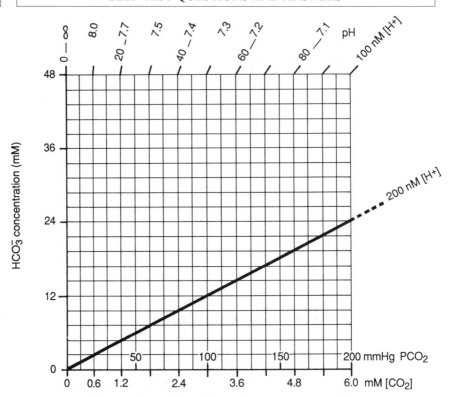

Figure 8.1. Figure accompanying structured exercise 1.

Construction of lines joining points of equal [H$^+$] (also called 'isohydric contours' or iso-pH lines).

1. From equation 1 and the given value of K, calculate [H$^+$] in nM at the point (PCO_2 = 40 mmHg, [HCO$_3$$^-$] = 24 mM); these are typical values for arterial blood plasma.

   You can convert from PCO_2 (mmHg) to mM from the lower x-axis of the graph (40 mmHg is equivalent to 1.2 mM CO_2).

2. Convince yourself that you get the same answer if the PCO_2 and [HCO$_3$] are changed by the same factor; consider the non-physiological point (PCO_2 = 80 mmHg, [HCO$_3$$^-$] = 48 mM).

   Plot the two points and then the complete contour for the [H$^+$] value obtained.

3. If you double the PCO_2 but keep the [HCO$_3$$^-$] constant, what happens to the [H$^+$]? The answer to this question will prompt you to add a second isohydric contour.

   By this stage, the construction lines along the top of the outline graph should indicate to you how to construct a fan of isohydric contours; do this now, labelling each contour with its [H$^+$].

Graphs for real fluids. We start by considering water and then progressively add the plasma constituents important in acid–base balance.

B. Distilled water

We now consider the effect of changes in P_{CO_2} on $[HCO_3^-]$ in distilled water. Every molecule of CO_2 which dissociates yields one H^+ ion and one HCO_3^- ion. There is no buffering of H^+ ions. So, neglecting the H^+ contributed by the dissociation of water, we have:

$$[H^+] = [HCO_3^-].$$

For water equilibrated with gas containing no CO_2, plot a point on the graph.

For water equilibrated with $P_{CO_2} = 120$ mmHg, calculate $[HCO_3^-]$ to the nearest power of 10 ($10^{-1}, 10^{-2} \ldots 10^{-6}$). Take care in work in M. Plot this point on your graph.

Sketch the $[HCO_3^-]$ against P_{CO_2} relation for distilled water and label it.

C. Sodium bicarbonate solution

We next calculate the effect of changes in P_{CO_2} for a solution of sodium bicarbonate. We choose a concentration of $[HCO_3^-] = 24$ mM since this is a typical concentration of bicarbonate for normal plasma. Plot the point representing $P_{CO_2} = 0$.

For a given P_{CO_2}, we could calculate the $[HCO_3^-]$, just as we did for water. However, the calculation is tedious and we adopt a more intuitive approach. Every molecule of CO_2 which dissociates yields one ion of H^+ and one of HCO_3^-. As with distilled water, there is no buffering of H^+ ions.

We wish to estimate the $[HCO_3^-]$ when the $P_{CO_2} = 200$ mmHg, i.e. $[CO_2] = 6$ mM. We know that $[HCO_3^-]$ must be at least 24 mM, since there was that much present in the solution before exposing it to an increased P_{CO_2}. Convince yourself that, as you move vertically upwards along the line representing $P_{CO_2} = 200$ mmHg, the $[H^+]$ is decreasing. Since $[HCO_3^-]$ must be at least 24 mM, the highest possible value for $[H^+]$ is 200 nM (from the isohydric contour for $[H^+] = 200$ nM). So the maximum possible **increase** in $[HCO_3^-]$ is also 200 nM or 0.0002 mM. The total $[HCO_3^-]$ is therefore not more than $(24$ mM $+ 0.0002$ mM$) = 24.0002$ mM.

You are now in a position to plot the point for $P_{CO_2} = 200$ mmHg. on the graph and hence to complete the $[HCO_3^-]$ against $[CO_2]$ relationship; do this and label the plot.

Conclusions Delete the inappropriate terms in the brackets below. For sodium bicarbonate solution, increasing the P_{CO_2} results in:

(large/insignificant) change of $[HCO_3^-]$ with
(large/insignificant) change of $[H^+]$.

The buffer pair CO_2-HCO_3^- is an (efficient/inefficient) buffer for changes in P_{CO_2}.

D. Perfect buffer

The perfect buffer (i.e. a solution whose $[H^+]$ is constant no matter what acid or alkali we add) does not exist, but let us assume that we can approximate to it. Suppose that such a solution holds the $[H^+]$ at 40 nM and that it contains 24 mM $NaHCO_3$ solution when the P_{CO_2} is 4 mmHg. Show on the graph the $[HCO_3^-]$ against P_{CO_2} relationship for this solution and label it.

E. Blood plasma

This contains, amongst other chemicals, bicarbonate and the non-bicarbonate buffers plasma protein and phosphates. (Quantitatively, the plasma proteins are much more important than the phosphates in buffering).

These buffers have a limited buffering capacity so that the $[HCO_3^-]$ against P_{CO_2} curve for plasma is intermediate between those which you have drawn in Sections C and D.

An average $[HCO_3^-]$ for plasma is 24 mM when in equilibrium with P_{CO_2} of 40 mmHg. Plot this point and make an informed guess of the relationship on either side of this point; extend your plot to cover the P_{CO_2} range from zero to 100 mmHg. Now draw graphs to indicate the effects of increase or decrease in the (non-bicarbonate) buffering power of the solution.

F. Blood

The main non-bicarbonate buffer in whole blood is haemoglobin. Sketch on your graph the difference between the plasma $[HCO_3^-]$ against P_{CO_2} relationships comparing whole blood and plasma separated from its erythrocytes.

CHAPTER 3

Question 1 (MCQ)

Concerning uncompensated respiratory acidosis:

A. This condition is caused by hyperventilation.
B. Respiratory movements may be vigorous.
C. The rise in arterial $[H^+]$ is in direct proportion to the rise in arterial P_{CO_2}.
D. The plasma $[HCO_3^-]$ is normal.
E. If an arterial blood sample is equilibrated with a gas containing CO_2 at a tension of 40 mmHg, the pH of the sample is normal.

Question 2 (MCQ)

By comparison with the situation in a healthy person breathing normally, immediately after the onset of hypoventilation (as in a severe attack of asthma):

A. The alveolar P_{CO_2} falls.
B. The amount of carbon dioxide in physical solution in the blood changes in direct proportion to the change in the partial pressure of carbon dioxide.
C. The rate of excretion of carbon dioxide in expired gas is lower than the rate of CO_2 production in the tissues of the body.
D. The alveolar P_{O_2} falls.
E. The arterial concentration of oxygen falls.

Question 3 (MCQ)

In fully compensated respiratory acidosis:

A. The alveolar P_{CO_2} is high.
B. The plasma $[HCO_3^-]$ is high.
C. The plasma $[H^+]$ is high.
D. Compensation involves hyperventilation.
E. Compensation is complete typically within six hours of the onset of hypoventilation.

Question 4 (MCQ)

With regard to acid–base physiology:

A. Human urine is always acid.
B. In disorders of acid–base status, the concentrations of potassium and hydrogen ions in the extracellular fluid tend to move together in the same direction.
C. Respiratory compensation in metabolic alkalosis involves a fall in arterial P_{CO_2}.
D. Compensation of respiratory acidosis involves increased renal excretion of bicarbonate ions.
E. If in the compensated phase of respiratory acidosis, a sample of arterial blood is equilibrated with a gas mixture containing CO_2 at a partial pressure of 40 mmHg, the pH of the sample is normal.

Question 5 (MCQ)

In a patient with metabolic acidosis:

A. The condition may be caused by vomiting of gastric contents.
B. The $[HCO_3^-]$ of arterial blood is raised.
C. The urine is alkaline.
D. The kidney retains bicarbonate.
E. There is likely to be hyperventilation.

Question 6 (MCQ)

Compared with the graph relating the $[HCO_3^-]$ as a function of P_{CO_2} for normal arterial blood:

A. For anaemic blood the slope of the graph is greater.
B. For arterial blood from a subject with acute respiratory acidosis, the graph lies higher.
C. For arterial blood from a subject with acute metabolic acidosis, the graph lies higher.
D. If to normal blood some sodium bicarbonate were added, the graph would lie higher.
E. If to normal blood some sodium hydroxide were added, the graph would lie higher.

Question 7 (MCQ)

With regard to buffering:

A. In plasma, phosphate is quantitatively more important than other non-bicarbonate buffers.
B. Interstitial fluid in an inflamed area is better buffered than that in non-inflamed tissue.
C. The CO_2–bicarbonate buffer system is an efficient buffer system in a closed system.
D. The CO_2–bicarbonate buffer system is an efficient physiological buffer system.
E. A buffer pair is most effective as a buffer at pH values far from their pK value.

Question 8 (MCQ)

Lowered plasma potassium:

A. tends to decrease renal H^+ secretion;
B. is commonly associated with respiratory alkalosis;
C. is commonly associated with metabolic alkalosis;
D. is a cause of metabolic alkalosis;
E. is commonly associated with an intracellular alkalosis.

CHAPTER 4

Structured exercise 2

Review of material in Chapters 3 and 4.

The exercise consists of working through the text, filling in the blank spaces and deleting inappropriate expressions enclosed in brackets. In brackets marked*, more than one entry may be correct. As you proceed through exercise A, you should label the head of each arrow on your diagram with the condition which it represents.

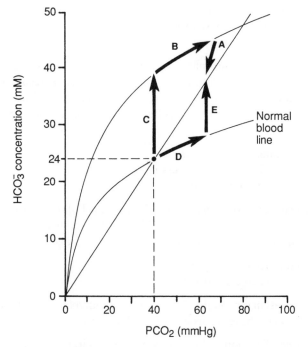

Figure 8.2. Figure accompanying structured exercise 2.

A. Acidosis and alkalosis

1. Study Figure 8.2, a diagrammatic representation of acid–base status; the labelling of arrows is random.

2. **Respiratory acidosis** Hypoventilation of acute onset results in a change of blood biochemistry (along the normal, to a different) blood line, as shown by arrow ... In the resultant uncompensated respiratory acidosis the alveolar P_{CO_2} is

(high, near normal, low); the alveolar P_{O_2} is
(high, near normal, low); $[H^+]$ is
(high, near normal, low); base excess is
(positive, close to zero, negative).

Renal compensation for persisting hypoventilation is brought about by the *(generation, retention, excretion) of HCO_3^- ions; the change in acid–base status is indicated by arrow ... The compensation has, as a result, the restoration towards normal of (P_{CO_2}, $[HCO_3^-]$, $[H^+]$).

In the resultant compensated respiratory acidosis, by comparison with normal, the P_{CO_2} is

(high, near normal, low); $[H^+]$ is
(high, near normal, low); base excess is
(positive, about zero, negative).

3. **Metabolic alkalosis** This may be due to *(ingestion of NH_4Cl, ingestion of sodium citrate, persistent vomiting of gastric contents). It results in a change in blood biochemistry (along the same, to a different) blood line, as shown by arrow ... At the stage when the resultant disturbance is uncompensated, the P_{CO_2} is

(high, near normal, low); $[H^+]$ is
(high, near normal, low); base excess is
(positive, close to zero, negative).

Respiratory compensation results in a change in $[H^+]$ (back towards, further away from) normal and a change in $[HCO_3^-]$ (back towards, further away from) normal. It is represented by arrow ... At this stage the P_{CO_2} is

(high, around normal, low); base excess is
(positive, close to zero, negative).

Renal compensation is brought about by (excretion, retention) of HCO_3^- ions; the change in acid–base status is indicated by arrow ....

4. **Comparison.** The points for compensated respiratory acidosis and compensated metabolic alkalosis are (much closer together than, further apart than) the points representing the uncompensated disorders. Consequently, it is (easy, difficult) from the biochemistry of the blood alone to discriminate between the compensated disorders.

5. **Conclusion.** It is (relatively easy, difficult) from the examination of the blood alone to identify uncompensated disorders of acid–base balance but (relatively easy, difficult) to identify compensated disorders without ambiguity. From a study of the blood alone, (uncompensated, compensated) disorders of acid–base balance cannot be identified; each separate factor must be sought and its effect evaluated.

B. Disturbances of acid–base balance and of fluid volume: interactions

Vomiting of gastric contents. A common example occurs in the vomiting of gastric contents, which results in the loss of hydrogen ions. When the vomiting is mild to moderate, the metabolic alkalosis initiates the sequence of events which you followed in section A3.

With persistent prolonged vomiting of gastric contents the loss of water and electrolytes becomes severe. The response to dehydration takes precedence over acid–base adjustments. This may lead to the condition of **paradoxical aciduria**. The remainder of this exercise conducts you through one of the mechanisms contributing to this situation.

Persistent vomiting of gastric contents. There is a loss of HCl, NaCl, water. Hence the kidney is called upon to retain sodium and chloride and to restore fluid volume as well as respond to the disturbance of acid–base balance.

Start by considering the renal tubular handling of ions in health. Sodium and chloride ions are readily filtered at the glomerulus. The plasma concentrations are: sodium .... mM, chloride .... mM. Most of the filtered load of sodium and chloride is reabsorbed.

Draw a diagram as you proceed. Indicate the movement of ions from renal tubular fluid to interstitial fluid. Draw arrows to indicate the movements of sodium and chloride ions, making the arrows roughly proportional in length to the amount of the ions reabsorbed.

Electrochemical neutrality must be maintained; for every mole of Na^+ reabsorbed, its charge must be balanced by an equivalent transfer of other ions. This is achieved partly by movement of ....-charged ions from urine to renal interstitial fluid (ISF) and partly by movement of ....- charged ions from ISF to urine. The ions involved include Cl^-, HCO_3^-, H^+, K^+. Indicate these on your diagram, with arrows whose directions are appropriate to make up for the fact that the chloride arrow is shorter than the sodium arrow.

Now we consider how the diagram is modified in a patient with persistent vomiting of gastric contents. Because of the loss of hydrochloric acid in the vomitus, the concentration of chloride in the plasma (falls, rises) and the amount of chloride filtered at the glomerulus is (more, less) than normal. The difference between the lengths of the chloride and sodium arrows is therefore (reduced, exacerbated). Consequently there will tend to be an (increase, decrease) in movement of HCO_3^-, H^+, K^+ through the tubule wall in the directions of the relevant arrows. This will result in

(retention, excretion) of HCO_3^-;
(retention, excretion) of H^+;
(retention, excretion) of K^+.

Acid–base. So, despite an extracellular metabolic alkalosis, the urine is (acid, alkaline) and (free of, loaded with) HCO_3^-. This (helps to compensate for, exacerbates) the metabolic alkalosis; it is a (positive, negative) feedback system.

Potassium. Draw a diagram as you proceed through this paragraph. As a result of the renal effects, the body is (depleted of, loaded with) potassium and the plasma potassium is (high, low). Consequently, potassium will tend to (leave, enter) the cells of the body. To maintain electrochemical neutrality, this will lead to hydrogen ions (entering, leaving) the cells so that the extracellular-metabolic alkalosis is (exacerbated, returned towards normal). There is an intracellular (acidosis, alkalosis).

Endocrine factors add to this effect. A low plasma chloride concentration promotes aldosterone secretion which, in turn, promotes sodium reabsorption and renal loss of potassium and hydrogen ions.

Conclusion. The acidity of the urine is (appropriate, inappropriate) as a response to metabolic alkalosis. The preservation of electrolyte and fluid volume (takes precedence, is subservient) to the correction of the acid–base disturbance. Cautious intravenous infusion of isotonic sodium chloride solution is likely to (improve, have no effect on, adversely affect) the patient's condition.

Question 9 (MCQ)

In human plasma:

A. The concentration of CO_2 exceeds that of HCO_3^-.
B. At constant P_{CO_2}, pH falls with increasing bicarbonate concentration.
C. Base excess is near zero in a healthy person.
D. Base excess is near zero in plasma from an anaemic patient.
E. The total buffer base is low in an anaemic person.

Question 10 (MCQ)

In the arterial blood of a patient who is anaemic but otherwise well:

A. The P_{CO_2} is normal.
B. The $[HCO_3^-]$ is normal.
C. The $[H^+]$ is normal.
D. The non-bicarbonate buffer base is normal.
E. The base excess is close to zero.

Question 11 (MCQ)

Doubling of the partial pressure of CO_2 to which a sample of blood *in vitro* is equilibrated will result in:

A. a doubling of the concentration of CO_2 in physical solution;
B. a doubling of $[H_2CO_3]$;
C. a doubling of $[HCO_3^-]$;
D. a decrease in $[H^+]$;
E. a decrease in the concentration of non-bicarbonate buffer base.

The next four questions refer to Figure 8.3.

Question 12 (MCQ)

Given that the point N represents the values found in a sample of arterial blood taken from a healthy person and analysed immediately:

A. a typical value for x would be 80 mmHg;
B. a typical value for Y would be 24 mM;

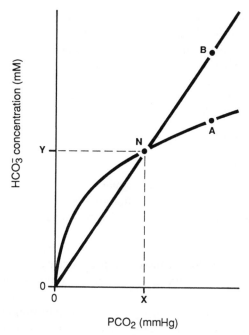

Figure 8.3. Figure accompanying MCQ questions 12 to 15.

C. the oblique line through the origin joins points of increasing $[H^+]$;
D. blood represented by point B would have a positive base excess;
E. blood represented by point A would have a normal standard bicarbonate.

Question 13 (MCQ)

A patient whose acid–base status is originally normal suddenly suffers from hypoventilation. In this patient:

A. Within half an hour, the point A could represent the acid–base status of the arterial blood.
B. Within half an hour, the hydrogen ion concentration of the blood is likely to have fallen.
C. After half an hour, the base excess would be near zero.
D. After a few days with persistent hypoventilation, the acid–base status of the blood might be represented by a point close to B.
E. After a few days with persistent hypoventilation, the hydrogen ion concentration of the blood is likely to be nearer to normal than half an hour after the onset of the hypoventilation.

Question 14 (MCQ)

With further reference to the figure, and possible values which might be found in the arterial blood of different patients:

A. Point A unambiguously indicates the disorder as being primarily respiratory.
B. Point B unambiguously indicates the disorder as being primarily metabolic.
C. Blood from a patient with an uncompensated respiratory alkalosis would yield a point lying on the 'normal blood line'.
D. Blood from a patient with an uncompensated metabolic acidosis would yield a point lying on the 'normal blood line'.
E. At point B, the total buffer base is above normal.

Question 15 (MCQ)

With further reference to the figure:

A. At N, the $[H^+]$ and $[OH^-]$ are equal.
B. Ingestion of a large amount of ammonium chloride would result in a change which would be represented by moving along the 'normal blood line'.
C. In anaemia, the whole of the blood line would be displaced downwards.
D. If a sample blood whose chemistry is represented by point A is equilibrated in vitro with a gas containing CO_2 at a tension of 40 mmHg, the pH of the sample is normal.
E. If a sample blood whose chemistry is represented by point B is equilibrated *in vitro* with a gas containing CO_2 at a tension of 40 mmHg, the pH of the sample is normal.

Structured exercise 3

Table 8.1 shows determinations on samples of arterial blood from four untreated cases. For each, blood was taken soon after the onset of the disorder and then again a few days later. For each case, identify the acid–base status at each of

Table 8.1

Case	Soon after onset of disorder or after a few days	pH	P_{CO_2} mmHg	Base Excess mM
1	onset	7.20	80	0
1	later	7.35	80	+10
2	onset	7.5	28	0
2	later	7.4	28	−6
3	onset	7.05	41	−19
3	later	7.27	23	−15
4	onset	7.60	40	+15
4	later	7.55	54	+15

the two stages and explain the changes which occurred between the first and second analyses.

CHAPTER 5

Question 16 (MCQ)

The following tend to cause metabolic acidosis:

A. ingested methionine;
B. sodium lactate in haemodialysis fluids;
C. a diet of meat;
D. strenuous exercise;
E. secretion of gastric acid.

CHAPTER 6

Question 17 (MCQ)

As blood passes through a systemic capillary:

A. Chloride ions move from the erythrocyte cytoplasm to the plasma.
B. The erythrocytes shrink.
C. The blood becomes more alkaline.
D. The total buffer base within the erythrocytes falls.
E. The total buffer base in the plasma falls.

CHAPTERS 1 AND 6

Structured exercise 4 Partial pressures and concentrations of gases

For this exercise, you are to take atmospheric pressure as being 760 mmHg. You are also given that the solubility coefficient of oxygen is 0.03 ml O_2 per mmHg per litre of aqueous solution.

Give numerical answers to three significant figures.

4.1A. If a sample of alveolar gas is at atmospheric pressure and contains 14% oxygen, what is the partial pressure of oxygen in the gas?
4.1B. If a sample of alveolar gas is at atmospheric pressure and contains carbon dioxide at a partial pressure of 40 mmHg, what is the concentration of carbon dioxide in the gas?
4.1C. Draw a graph to show the relationship between the concentration of oxygen (on the y-axis) as a function of the partial pressure of oxygen

(on the x-axis) in (a) a gas mixture, (b) physical solution in an aqueous medium and (c) normal whole blood.

Definitions

4.2A. Define the partial pressure of a gas, giving the units in which it is measured.

4.2B. Define the concentration of a gas, giving the units in which it may be measured.

4.3. For gas concentrations, there are three situations to consider:
  i) a gas in a gas mixture;
  ii) a gas in physical solution in a liquid;
  iii) a gas which reacts chemically with chemicals in the liquid.

To which of these situations do the following pertain? In each case, explain your answer.

4.3A. The concentration as a percentage is the same as the partial pressure as a percentage of the total pressure.

4.3B. The concentration is directly proportional to the partial pressure.

4.3C. The coefficient of solubility is necessary and sufficient to allow the calculation of concentration from the partial pressure.

4.4. **Numerical examples**: given the solubility coefficient of oxygen as 0.03 ml O_2 per mmHg per litre of aqueous solution and the oxygen dissociation curve of Figure 6.1B, calculate the following. In each calculation, indicate your working clearly.

4.4A. the partial pressure of oxygen in air compressed to a pressure of two atmospheres, given that the air contains 20% oxygen;

4.4B. the concentration of oxygen in expired gas at one atmosphere (760 mmHg), the oxygen partial pressure being 120 mmHg. (express your answer as the fractional concentration and as a percentage of total);

4.4C. the concentration of oxygen in physical solution in blood at an oxygen partial pressure of 100 mmHg;

4.4D. the concentration of oxygen in chemical combination with haemoglobin in blood at an oxygen partial pressure of 40 mmHg;

4.4E. at an oxygen partial pressure of 40 mmHg, the amount of oxygen present in blood in physical solution as a percentage of the amount combined with haemoglobin.

Question 18 (MCQ)

In one bowl you place water and in another you place blood; you expose both fluids to the air and allow time for equilibration.

A. The partial pressure of oxygen in the air equals that in the water.

B. The partial pressure of oxygen in the air equals that in the blood.
C. The partial pressure of oxygen in the water equals that in the blood.
D. The total concentration of oxygen in the blood equals that in the air.
E. The concentration of oxygen in physical solution in the water equals that in physical solution in the blood.

Question 19 (MCQ)

Concerning haemoglobin:

A. Oxyhaemoglobin contains iron in the Fe^{3+} (ferric) form.
B. Oxyhaemoglobin is a stronger acid than haemoglobin.
C. The affinity of haemoglobin for oxygen is increased by an increase in the P_{CO_2}.
D. The affinity of haemoglobin for oxygen is increased by an increased hydrogen ion concentration.
E. The shift in the carbon dioxide dissociation curve by oxygenation of the blood is because of altered affinity of haemoglobin for hydrogen ions.

Carriage of carbon dioxide

Question 20 (MCQ)

With reference to carbon dioxide:

A. it is taken up more readily by desaturated than by fully oxygenated blood;
B. most of the CO_2 which can be chemically released from the blood is in the form of bicarbonate;
C. its diffusion into the erythrocytes is slow;
D. it reacts chemically more quickly in plasma than in erythrocytes;
E. about a quarter of the CO_2 taken up by the blood in the tissues is carried to the lungs as carbamino-haemoglobin.

Question 21 (MCQ)

By comparison with the state of affairs in a healthy person breathing normally, immediately after the onset of a period of hyperventilation:

A. the partial pressure of carbon dioxide in the alveolar gas rises;
B. the amount of carbon dioxide excreted per minute in the expired gas rises;
C. the rate of excretion of carbon dioxide in expired gas exceeds the rate of production in the body;
D. the partial pressure of oxygen in the alveolar gas rises;
E. the change in oxygen content of arterial blood is in direct proportion to the change in the partial pressure of oxygen.

CHAPTER 7

Structured exercise 5: Urine

5.1. Assuming that urine in the bladder has a P_{CO_2} 40 mmHg, from the Henderson-Hasselbalch equation calculate, for urines of the following pH values, the concentrations of urinary bicarbonate.

Urine pH   urinary $[HCO_3{}^-]$
7.4
6.4
5.4
4.4

5.2. The following values were obtained in a normal person:
Plasma $[HCO_3{}^-] = 24$ mM
Urine pH $= 6.4$
Glomerular filtration rate $= 120$ ml per min.
Urine flow $= 1$ ml per min.
Calculate:
  i) the amount of bicarbonate filtered per min.;
  ii) the amount of bicarbonate excreted in the urine per min.;
  iii) the percentage of filtered bicarbonate which is excreted in the urine.

5.3. From a normal adult person, the following values were obtained:
24-hour urine collection

Urine total volume	1.5 litres
Urine $[HCO_3{}^-]$	1 mM
Urine [ammonium]	30 mM
Urine [titratable acid]	16 mM
Plasma $[HCO_3{}^-]$	24 mM
Plasma [ammonium]	30 micromole/litre
Glomerular filtration rate	180 litres/24 hr

For this subject, calculate the following:

5.3A. the total acid excretion in mmole/24 hr;
5.3B. the percentage of filtered fluid which was excreted;
5.3C. the percentage of filtered bicarbonate which was excreted;
5.3D. the amount of bicarbonate (in mmole) reclaimed during the 24-hr period.
5.3E. the 'renal clearance' for ammonium ions
5.3F. Does the answer to 5.3E exceed the renal plasma flow? Explain your answer.

Question 22 (MCQ)

A patient has an arterial P_{CO_2} of 50 mmHg and a bicarbonate concentration of 15 mM. His kidneys are healthy.

A. The pH of his arterial plasma is less than 7.4.
B. Respiratory acidosis is a component of his acid–base disorder.
C. Metabolic acidosis is a component of his acid–base disorder.
D. His plasma $[K^+]$ is likely to be low as a result of the acid–base disturbance.
E. His renal glutaminase levels are likely to be low.

Question 23 (MCQ)

In a person it was found that on one occasion 0.01% of the filtered load of bicarbonate was excreted. (This is a typical value for a normal healthy person.) Six months later, this percentage had changed to 0.2%. Possible causes of this change are:

A. treatment with an inhibitor of carbonic anhydrase;
B. change to a diet high in meat content;
C. ingestion of a large amount of sodium acetate;
D. ingestion of a large amount of ammonium chloride;
E. chronic renal failure.

Question 24 (MCQ)

With reference to renal mechanisms:

A. Glucose molecules entering renal venous blood as a result of reabsorption in the proximal convoluted tubule are the same molecules that were in the tubular fluid.
B. Bicarbonate ions entering renal venous blood as a result of reabsorption in the proximal convoluted tubule are the same ions that were in the tubular fluid.
C. The proximal tubule reabsorbs more hydrogen ions than does the distal tubule.
D. The transport of hydrogen ions is against a greater gradient in the proximal tubule than in the distal tubule.
E. The change in hydrogen ion concentration as tubular fluid passes through the proximal convoluted tubule is greater than the change as the fluid passes through the distal convoluted tubule.

Question 25 (MCQ)

In a person excreting acid urine, as the tubular fluid passes along the nephron:
A. there are regions of the nephron in which the hydrogen ion concentration is less than in the plasma;
B. the reabsorption of water assists in the excretion of hydrogen ions;
C. the reabsorption of water contributes to the rise in hydrogen ion concentration;
D. the lower limit of urinary acidity is around 7.0 pH units;
E. in the distal convoluted tubule, hydrogen ions flowing from within the tubular cell cytoplasm to the tubular fluid are moving down their electrochemical gradient.

Question 26 (MCQ)

Concerning the role of ammonia in renal function:

A. ammonium ions in the tubular fluid are derived from the glomerular filtrate;
B. the renal clearance of ammonium ions may exceed the renal plasma flow;
C. the contribution of ammonia to excretion of acid is included in the 'titratable acid' in the urine;
D. ammonium ions cross cell membranes more readily than ammonia;
E. ammonia is only produced in the distal nephron.

Question 27 (MCQ)

When the P_{CO_2} increases:
A. the bicarbonate concentration rises more in interstitial fluid *in vitro* than in plasma *in vitro*;
B. *in vivo*, bicarbonate moves from the blood to the interstitial fluid;
C. the ratio of $H_2PO_4^-$ to HPO_4^{2-} in plasma increases;
D. H^+ secretion in the kidney rises;
E. the excretion of ammonium ions in the urine increases.

Essay question

A person hyperventilated steadily for several days.
Consider these variables in the arterial blood which were initially at their normal values:

(a) the partial pressure of carbon dioxide;
(b) the partial pressure and content of oxygen;
(c) the plasma concentration of hydrogen ions (in terms of either nM or pH);
(d) the plasma concentration of bicarbonate ions.

For each variable give a typical normal value, and likely values during hyperventilation; for (c) and (d) distinguish between the changes half an hour after the onset of hyperventilation and those which develop over several days.

Structured question

A. Write an equation relating:
   i) the concentration of hydrogen ions or the pH;
   ii) the concentration or the partial pressure of carbon dioxide; and
   iii) the concentration of bicarbonate ions in an aqueous solution.
B. Define 'metabolic acidosis'.
C. Give an example of occurrence of metabolic acidosis.
D. Draw a graph to show the inter-relation of these factors for 'separated plasma' and for 'true plasma' (as in whole blood). Label each axis to

indicate the chemical species and add scales. Indicate the coordinates of a point corresponding to normal arterial blood.

E. Consider the changes in blood chemistry in metabolic acidosis; show on your graph for 'true plasma' and explain below the effects of:
  i) the initial addition acid;
  ii) the homeostatic mechanisms which come into play.

ANSWERS TO SELF-TEST

CHAPTERS 1 AND 2

Key to structured exercise 1: Acid–base status—introduction to its graphic representation

A. Derivation of Henderson-Hasselbalch equation

$$CO_2 + H_2O = H^+ + HCO_3^-$$

By the law of mass action, the rate of forward reaction is proportional to $[CO_2] \times [H_2O]$ and the rate of backward reaction is proportional to $[H^+] \times [HCO_3^-]$.

At equilibrium the rate of forward reaction equals the rate of backward reaction. Since the concentration of water $[H_2O]$ is approximately constant, we can incorporate this factor in the constant of proportionality to yield

$$[H^+][HCO_3^-] = K[CO_2]$$
$$[H^+] = K[CO_2]/[HCO_3^-] \text{ (equation 1)}$$
$$[H^+] = 0.8 \times 10^{-6}[CO_2]/[HCO_3^-]$$

Dimensions of K: From equation 1, $[CO_2]/HCO_3^-]$ is a ratio and therefore dimensionless. Hence $[H^+]$ is proportional to K; since $[H^+]$ is a concentration, K must have the dimension of concentration too, i.e. K has the dimensions of M.

A. It is possible to label the x-axis either as P_{CO_2} or as $[CO_2]$ since $[CO_2]$ is directly proportional to P_{CO_2}.

Calculation

A1. $[H^+] = 0.8 \times 10^{-6}\left(\dfrac{1.2 \times 10^{-3}}{24 \times 10^{-3}}\right) = 0.4 \times 10^{-7}$ M. or 40×10^{-9} M or 40 nM.

2. Ditto.

3. $[H^+] = 80$ nM.

B. Distilled water
When no CO_2 is present, $[HCO_3^-] = $ zero.

(a) This gives the point (0.0) as being on the graph.

(b) Every molecule of CO_2 which dissociates yields one ion of H^+ and one of HCO_3^-. There is no other buffer to remove H^+. So neglecting the contribution of H^+ from the dissociation of water:

$$[H^+] = [HCO_3^-].$$

From equation 1:

$$[H^+] = 0.8 \times 10^{-6}[CO_2]/HCO_3^-]$$
When $[H^+] = [HCO_3^-]$, then $[HCO_3^-]^2 = 0.8 \times 10^{-6}[CO_2]$.
With $[CO_2] = 3.6 \times 10^{-3}$ M,
$$[HCO_3^-]2 = 0.8 \times 10^{-6} \times 3.6 \times 10^{-3} < 10^{-8}.$$

$[HCO_3^-] < 10^{-4}$ M or 1/10th mM,. i.e. less than one-tenth mM. This cannot be distinguished from zero on the graph. It gives the point (120, approx 0) on the graph. Hence the distilled water line is virtually the x-axis.

C. Sodium bicarbonate solution 24 mM
$NaHCO_3$ is a salt and so completely dissociates in water.
$NaHCO_3$ dissociates to $[Na^+] = 24$ mM and $[HCO_3^-] = 24$ mM
So for $PCO_2 = $ zero, $[HCO_3^-] = 24$ mM, which is plotted as the point (0,24).

For $PCO_2 = 200$ mmHg $[HCO_3^-] = 24$ mM plus a contribution from CO_2 dissociation. Each gram molecule (mole) of CO_2 which dissociates yields one mole of H^+ and one of HCO_3^-.
We can estimate how much $[H^+]$ has been released. Since $[HCO_3^-]$ must be at least 24 mM, the highest possible value for $[H^+]$ is 200 nM (from isohydric contours). Any increase in value of $[HCO_3^-]$ involves a fall in $[H^+]$. So a maximum value for increase in $[HCO_3^-]$ is **200 nM** or **0.0002 mM**. So the total $[HCO_3^-] = 24.0002$ mM. On the graph, this cannot be resolved from 24 mM. So, to a very close approximation, the point (200 mmHg, 24 mM) is on the bicarbonate line. The sodium bicarbonate solution line lies almost on a horizontal line.
This indicates a total lack of buffering. As the PCO_2 rises, the line crosses isohydric contours as quickly as it can and the $[H^+]$ rises rapidly.

D. $NaHCO_3$ solution plus non-bicarbonate buffer
If the buffer is near as possible perfect, then $[H^+] = 40$ nM and the buffer line is the isohydric contour.

E. Blood plasma is not a very good buffer.

F. Whole blood is a much better buffer.

CHAPTER 3

Answer 1

A. No. It is hypoventilation.
B. Yes. If the cause is respiratory obstruction.
C. No. The rise in $[H^+]$ is buffered by blood buffers.
D. No. It is higher than normal.
E. Yes. The abnormality is primarily an increase in P_{CO_2} and if this is reversed the blood chemistry is normal.

Answer 2

A. No. It rises.
B. Yes. It is a principle of physical chemistry that there is always a direct proportionality between the partial pressure of a gas and its concentration in physical solution.
C. Yes. There is retention of CO_2 in the body as the arterial P_{CO_2} gradually rises.
D. Yes.
E. Yes. The arterial blood is no longer saturated with oxygen.

Answer 3

A. Yes. This is the primary abnormality.
B. Yes, both because of the high P_{CO_2} and because of renal retention of bicarbonate in this condition.
C. No. Full compensation implies that the plasma $[H^+]$ is normal.
D. No. Ventilation is compromised.
E. No. It takes days.

Answer 4

A. No. In alkalosis, the urinary pH can rise to eight units.
B. Yes. The mechanism is described in the text.
C. No. Alkalosis depresses respiration, so the arterial P_{CO_2} rises.
D. No. Bicarbonate is retained in order to return the blood pH towards normal and acid urine is excreted.
E. No. In this situation there is renal retention of bicarbonate so that, with a normal P_{CO_2}, the blood is more alkaline than usual.

Answer 5

A. No. This would cause alkalosis.
B. No. The $[HCO_3^-]$ is typically very low.
C. No. The kidney excretes an acid urine to rid the body of the unwanted acid.
D. Yes. This helps to buffer the acid load.
E. Yes. The peripheral chemoreceptors are stimulated by the raised arterial $[H^+]$.

Answer 6

A. No. The slope is less because anaemic blood is poorer than normal blood as a buffer.
B. No. The graphs would be coincident because acute respiratory acidosis means exposure of the subject to a high Pco_2 with no compensation.
C. No. The graph would lie below the normal because of excess non-respiratory acid.
D. Yes. This addition of extra bicarbonate would push the curve up.
E. Yes. The hydroxide ions react with CO_2 to provide the blood with more bicarbonate and so the effect on the blood line is the same as in D.

Answer 7

A. No. In plasma phosphate is a relatively unimportant buffer quantitatively because of its relatively low concentration.
B. Yes. In inflamed tissue, the capillary lining allows the passage of proteins, which act as good buffers.
C. No. The pK (6.1) is more than one pH unit from body pH (7.4).
D. Yes, because both Pco_2 and $[HCO_3{}^-]$ are under physiological control.
E. No. They buffer most effectively at pH values at the pK.

Answer 8

A. No. Low extracellular $[K^+]$ results in K^+ leaving renal tubular cells, H enters as part of an electroneutrality balance and the intracellular rise in $[H^+]$ assists renal H^+ secretion.
B. Yes. Hypokalaemia is commonly associated with alkalosis.
C. Yes, as in B.
D. Yes. By the mechanism described in A, H^+ moves from the extracellular compartment into the intracellular compartment.
E. No. By the mechanism described in A, there tends to be an intracellular acidosis.

CHAPTER 4

Key to structured exercise 2

A2. Respiratory acidosis
along the normal blood line, arrow D
high
low
high
close to zero

Renal compensation
generation and retention of bicarbonate ions, arrow E
$[H^+]$
high
near normal
positive

A3. Metabolic alkalosis
ingestion of sodium citrate, persistent vomiting of gastric contents
to a different blood line, arrow C
near normal
low
positive

Respiratory compensation
back towards
further away from, arrow B
high
positive

Renal compensation
excretion, arrow A

A4. Comparison
closer together
difficult

A5. Conclusion
relatively easy
difficult
compensated

B2. Persistent vomiting of gastric contents
sodium: 140 mM
chloride: 100 mM
negatively
positively
falls
less
exacerbated
increase
retention
excretion
excretion

B3. Acid–base
acid
free

exacerbates
positive

B4. Potassium
depleted
low
leave
entering
exacerbated
acidosis

B5. Conclusion
inappropriate
takes precedence
improve, because it provides electrolyte for the maintenance of extracellular
volume thus taking the pressure off mechanisms for electrolyte retention and
freeing mechanisms for acid–base homeostasis, which then correct the acid–
base disorder.

Answer 9

A. No. $[CO_2]$ is typically 2 mM and $[HCO_3{}^-]$ typically 24 mM.
B. No. The opposite.
C. Yes.
D. Yes. An anaemic person operates with arterial plasma at a normal P_{CO_2}, $[HCO_3{}^-]$ and pH; the base excess is therefore zero.
E. Yes, because the reduction in haemoglobin concentration results in a reduced buffering power of the blood.

Answer 10

A to C. Yes. The respiratory system holds the P_{CO_2} at a normal level, the $[HCO_3{}^-]$ is normal and so, by the Henderson-Hasselbalch equation, the $[H^+]$ is also normal.
D. No. The non-bicarbonate buffer base is mainly in the form of haemoglobin, the concentration of which is reduced in anaemia.
E. Yes. See A to C.

Answer 11

A. Yes. The amount of a gas in physical solution is directly proportional to the partial pressure of the gas with which the liquid is in equilibrium.
B. Yes. $H_2O + CO_2 = H_2CO_3$
   $[H_2O]$ is constant so by the law of mass action $[H_2CO_3]$ is directly proportional to $[CO_2]$.
C. No. This would imply that blood is a perfect buffer.

D. No. $[H^+]$ rises.
E. Yes. Most of the released hydrogen ions react with non-bicarbonate buffer base to yield non-bicarbonate buffer acid.

Answer 12

A. No. 40 would be a typical value.
B. Yes.
C. No. This line joins points of equal $[H^+]$.
D. Yes. It lies above the 'normal blood line'.
E. Yes. It lies on the 'normal blood line'.

Answer 13

A. Yes. This is uncompensated respiratory acidosis.
B. No. $[H^+]$ would have risen.
C. Yes. Renal compensation has not as yet had a chance to influence the blood biochemistry significantly.
D. Yes. The move vertically upward on this plot indicates renal retention of bicarbonate.
E. Yes. The renal retention of bicarbonate compensates for the change in $[H^+]$.

Answer 14

A. Yes. At the uncompensated stage, a move along the blood line to the right unambiguously indicates a primary uncompensated respiratory acidosis.
B. No. At point B, there are both respiratory and metabolic components.
C. Yes. It is the hallmark of uncompensated respiratory disorders that values from arterial blood give a point lying on the 'normal blood line'.
D. No. The point would lie below the normal blood line.
E. Yes. Any point above the 'normal blood line' has an increased value of total buffer base.

Answer 15

A. No. $[OH^-]$ exceeds $[H^+]$ several-fold.
B. No. This produces a metabolic acidosis; the point representing the subject's arterial blood moves down from the normal blood line.
C. No. The blood line for anaemic blood is less steep than for normal blood and the two curves intersect at $P_{CO_2} = 40\,mmHg$.
D. Yes, because the point represents blood on the normal blood line.
E. No, because the point representing the blood is above the normal blood line, so reducing the P_{CO_2} to $40\,mmHg$ brings the blood along its blood line which is higher than normal. So the $[H^+]$ at a P_{CO_2} of $40\,mmHg$ would be below normal.

Key to structured exercise 3

Case 1, onset: high P_{CO_2}, B.E. zero, uncompensated respiratory acidosis. Later, P_{CO_2} still high, B.E. positive: compensated respiratory acidosis.

Case 2, onset: low P_{CO_2}, B.E. zero, uncompensated respiratory alkalosis. Later, P_{CO_2} still low, B.E. negative: compensated respiratory alkalosis.

Case 3, onset: low pH, P_{CO_2} normal, negative B.E., uncompensated metabolic acidosis. Later: P_{CO_2} now low (hyperventilation) negative B.E. but less negative than before: metabolic acidosis with respiratory and renal compensation.

Case 4, onset: high pH, P_{CO_2} normal, positive B.E., uncompensated metabolic alkalosis. Later: P_{CO_2} now high (hypoventilation), B.E. unaltered: metabolic acidosis with respiratory compensation, no renal compensation.

CHAPTER 5

Answer 16

A. Yes. Methionine is a sulphur-containing amino acid; it is the most important source of labile methyl groups in normal conditions. Its metabolism yields sulphuric acid.
B. No. The lactate is metabolized so that effectively sodium hydroxide is being ingested.
C. Yes. Meat is rich in amino acids containing sulphur.
D. Yes. Strenuous exercise is accomplished by means of anaerobic metabolism; lactic acid is released from the active muscles into the blood stream.
E. No. Secretion of acid into the lumen of the stomach leaves the gastric venous blood with a raised concentration of bicarbonate, hence the slight 'alkaline tide' after feeding.

CHAPTER 6

Answer 17

A. Yes. This is the 'chloride shift' described in the text.
B. No. They swell due to an increase in total intracellular osmolar activity.
C. No. If the RQ is more than 0.7, the blood becomes more acid; if the RQ is 0.7, the reaction of the blood is unaltered.
D. Yes, because of loss of bicarbonate.
E. No. It rises because of gain of bicarbonate.

CHAPTERS 1 AND 6

Key to structured exercise 4 Partial pressures and concentrations of gases

4.1A. $\dfrac{14}{100} \times 760 = 106.4 \, \text{mmHg}$

4.1B. $\dfrac{40}{760} \times 100 = 5.2\%$

4.1C. (a) Gas mixture. At P_{O_2} of 100 mmHg, one litre of gas contains

$$\dfrac{100}{760} \times 1000 = 132 \, \text{ml} \, O_2$$

This gives point A in Figure 8.4.

(b) In physical solution in an aqueous medium. From the solubility coefficient for oxygen for a partial pressure of 1 mmHg, an aqueous solution holds 0.03 ml O_2 per litre in physical solution.

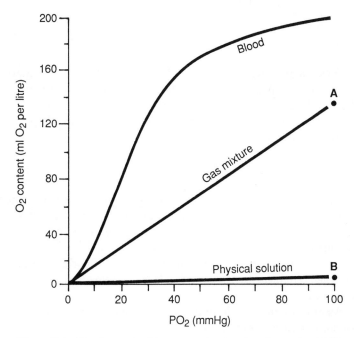

Figure 8.4. Graph required in answer to structured exercise 4.1C.

Therefore for a partial pressure of 100 mmHg, the amount is 3 ml O_2 per litre of solution.

This gives point B in Figure 8.4.

(c) This comes from Figure 6.1

4.2A. See Chapter 1, properties of gases, property 1.

4.2B. See Chapter 1, properties of gases, property 4.

4.3A. i) This follows from Chapter 1, properties of gases, property 4.

4.3B. i) and ii) This follows from Chapter 1, properties of gases, properties 4 and 5.

4.3C. ii) This follows from Chapter 1, properties of gases, property 5.

4.4A. $\dfrac{20}{100} \times (760 \times 2) = 304\,\text{mmHg}$

4.4B. $\dfrac{120}{760} = 0.158$ fractional content

15.8%

4.4C. $0.03 \times 100 = 3\,\text{ml}\ O_2$ per litre of blood

4.4D. $150\,\text{ml}\ O_2$ per litre of blood (graph 6.1)

4.4E. $\dfrac{0.03 \times 40}{150} \times 100 = 0.8\%$

Answer 18

A to C. Yes. Partial pressures are equal when equilibration is achieved.

D. Yes. By coincidence the blood contains typically 200 ml O_2 per litre, i.e. 20% and the air contains 20% oxygen.

E. Yes The concentration of a gas in physical solution is directly proportional to its partial pressure.

Answer 19

A. No. Iron is in the Fe^{2+} form in both oxyhaemoglobin and in haemoglobin; the formation of oxyhaemoglobin is not an oxidation reaction.

B. Yes. This helps to counteract the pH changes in blood caused by uptake and unloading of CO_2.

C. No. CO_2 shifts the oxygen dissociation curve to the right, i.e. the affinity for oxygen is reduced by CO_2 and this helps in unloading O_2 in the systemic capillaries.

D. No. As in C, an increase in hydrogen ion concentration shifts the oxygen dissociation curve to the right.

E. Yes. In deoxygenated blood, for instance, the haemoglobin accepts hydrogen ions and hence assists in the uptake of CO_2.

Carriage of carbon dioxide

Answer 20

A. Yes. This is a physiological mechanism to load blood with CO_2 with minimal change in P_{CO_2}.
B. Yes, about 85%.
C. No. CO_2 diffuses readily through cell membranes.
D. No. The opposite, because of the catalytic effect of carbonic anhydrase which is found in the erythrocytes but not in plasma.
E. Yes. This compound is responsible for a quarter of the arterio-venous difference in the CO_2 content of the blood.

Answer 21

A. No. The excess ventilation results in a dilution of alveolar gas with atmospheric air.
B. Yes. Carbon dioxide is washed out of the body.
C. Yes. See answer to B.
D. Yes. As in A, the alveolar gas is diluted with atmospheric air so the P_{O_2} of the alveolar gas rises.
E. No. The oxygen dissociation curve is not linear.

CHAPTER 7

Key to structured exercise 5 Urine

5.1. Urine pH $pH = 6.1 + \log\dfrac{[HCO_3^-]}{1.2}$

pH 7.4 $7.4 - 6.1 = \log\dfrac{[HCO_3^-]}{1.2}$

$[HCO_3^-] = 1.2 \times 10^{1.3} = 1.2 \times 20 = 24\,mM$

pH 6.4 $[HCO_3^-] = 1.2 \times 2 = 2.4\,mM$
pH 5.4 $[HCO_3^-] = 0.24\,mM$
pH 4.4 $[HCO_3^-] = 0.024\,mM$

5.2. i) Filtered bicarbonate $= \dfrac{120}{1000} \times 24 = 2.88\,mmole$ per min

ii) Excreted bicarbonate $= \dfrac{1}{1000} \times 2.4 = 0.0024\,mmole$ per min.

iii) $\dfrac{0.0024}{2.88} \times 100 = 0.08\%$ of filtered load is excreted

5.3A. $(16 + 30) \times 1.5 = 69$ mmole per 24 hr.

5.3B. $\dfrac{1.5}{180} \times 100 = 0.83\%$

5.3C. Bicarbonate filtered $= 24 \times 180 = 4320$ mmole per 24 hr
Bicarbonate excreted $= 1 \times 1.5 = 1.5$ mmole per 24 hr

Percentage excretion $= \dfrac{1.5}{4320} \times 100 = 0.035\%$

5.3D. Amount reclaimed $= 4320 - 1.5 = 4318.5$ mmole per 24 hr.

5.3E. $\dfrac{UV}{P} = \dfrac{30 \times 1.5}{30 \times 10^{-3}} = 1500$ litres per 24 hr

5.3E. Explanation.

The answer of 1500 litres a day exceeds the renal plasma flow (typically 900 litres a day). The reason is that the renal tubular cells are themselves a source of metabolically derived ammonia which diffuses into the tubular fluid and combines with hydrogen ions there to give ammonium ions. The renal clearance for ammonium therefore bears no relation to renal plasma flow. Clearance values of other chemicals are only indicative of such parameters as renal plasma flow (PAH) or glomerular filtration rate (inulin) for chemicals for which there is no source or sink in the kidneys themselves.

Answer 22

A. Yes. In the Henderson-Hasselbalch equation, with a low $[HCO_3{}^-]$ and a high $[CO_2]$, the pH is low on both accounts.
B. Yes, because his P_{CO_2} is high.
C. Yes, because $[HCO_3{}^-]$ is low in the presence of high P_{CO_2}; the point representing his blood must fall well below the 'normal blood line'.
D. No. In acidosis, the $[K^+]$ is if anything raised.
E. No. Glutaminase levels are likely to be high as part of the kidney's response to acidosis.

Answer 23

A. Yes, by inhibiting renal tubular H^+ secretion and hence reclaiming bicarbonate from the tubular fluid.
B. No. This would cause the urine to become more acid and its content of bicarbonate would fall.
C. Yes. This makes the body alkalotic.
D. No. The ammonium is removed by the liver leaving hydrochloric acid in the blood and making the body acidotic.
E. Yes. In chronic renal failure acidification of the urine is deficient and, as in A, bicarbonate is less adequately reclaimed from the tubular fluid.

Answer 24

A. Yes. The glucose is reabsorbed unaltered.
B. No. Retrieval of filtered bicarbonate depends on its chemical reaction with hydrogen ions to yield CO_2, diffusion of CO_2 into the tubular cell cytoplasm and recombination with water to yield hydrogen ions and bicarbonate again. See text.
C. Yes. See text.
D. No. The opposite is the case. See text.
E. Yes. See text.

Answer 25

A. No. There is a progressive fall in pH as the tubular fluid passes along the nephron.
B. No. Reabsorption of water is by mechanisms which do not assist excretion of hydrogen ions.
C. Yes. By removing water, the concentrations of all other chemicals in the tubular fluid, including hydrogen ions, are raised.
D. No. A correct answer would be 4.5 pH units; 7.0 would be a lower limit for the pH of the blood.
E. No. They are being pumped against an ever-increasing concentration gradient.

Answer 26

A. No. Ammonium ions are unique in being derived from ammonia which is synthesized in the tubular cells and diffuses thence into the tubular fluid.
B. Yes. Connected with the answer to A, the concentration of ammonium ions in the blood is minute, so a clearance calculation $C = (U \times V)/P$ may give a huge value, unrelated to any real flow as would be the case for the clearance of any other chemical.
C. No. The titratable acid is the acid buffered by buffers which enter the tubular fluid via the glomerulus.
D. No. The opposite is the case, this being the basis of the 'ammonium trap'.
E. No. Ammonia is produced along the whole length of both segments of the tubule.

Answer 27

A. No. The opposite because plasma is the better buffer.
B. Yes. Because blood is a good buffer, the bicarbonate concentration in blood rises and the bicarbonate diffuses thence into the interstitial fluid.
C. Yes. The reaction $HPO_4{}^{2-} + H^+ = H_2PO_4{}^-$ is driven to the right.
D. Yes. As a response to acidosis.
E. Yes, as for D; ammonium is carrying buffered hydrogen ions.

Key to essay question

Hyperventilation means overbreathing.

Partial pressures of O_2 and CO_2
Fall in arterial P_{CO_2}, rise in arterial P_{O_2}. Likely values: P_{CO_2} 40 to 25 mmHg. P_{O_2} 100 to 115 mmHg. Overbreathing causes alveolar gas to approach the composition of inspired air: there is a reduction in alveolar P_{CO_2}, and an increase in alveolar P_{O_2}. These same changes are reflected in the gas tensions in the blood leaving the lungs: therefore similar changes occur in the arterial blood. (Arterial P_{O_2} will be a little lower than alveolar P_{O_2}, as usual, because of small shunt and V/Q effects.)

The rate of removal of CO_2 from the blood initially exceeds the rate of its production by the tissues; the rate of uptake of oxygen by the blood is scarcely increased because, before hyperventilation, the haemoglobin is almost saturated with O_2. The blood CO_2 loss therefore initially exceeds oxygen uptake. When a new steady state is attained, CO_2 loss returns to normal, with alveolar P_{CO_2} low.

The changes in P_{O_2} and P_{CO_2} are comparable in magnitude; they are exactly the same if the RQ is 1.

O_2 content of arterial blood
Typical normal value is 200 ml per litre of blood. Because of the shape of the dissociation curve (show this), the rise in P_{O_2} is accompanied by very little increase in O_2 content, typically to 202 ml O_2 per litre of blood. (You can estimate this: arterial blood is usually about 97% saturated: the rise in P_{O_2} will increase this to about 98%, i.e. a 1% rise.)

Plasma hydrogen ion and bicarbonate concentrations
Typical normal values: $[H^+]$ 40 nM, $[HCO_3^-]$ 24 mM.

Half an hour after onset of hyperventilation, the $[HCO_3^-]$ has fallen but, since blood is not a perfect buffer, the fall in $[HCO_3^-]$ is not proportional to the fall in P_{CO_2}. Typical values after half an hour: $[H^+]$ 34 nM, $[HCO_3^-]$ 20 mM.

After three days, there is compensation (explain this) Typical values: $[H^+]$ 40 nM, $[HCO_3^-]$ 14 mM.

Appropriate equations and diagrams:

Oxygen dissociation curve.

Alveolar P_{O_2}, and alveolar P_{CO_2}, against alveolar ventilation.

Plasma $[HCO_3^-]$ against P_{CO_2}.

$$H_2O + CO_2 = H^+ + HCO_3^-$$

Henderson-Hasselbalch equation.

Key to structured question

a) $$pH = pK + \log_{10} \frac{[HCO_3^-]}{[CO_2]} \qquad pK = 6.1$$

$$\text{or} \quad pH = pK + \log_{10} \frac{[HCO_3^-]}{\alpha\, pCO_2 \quad \text{in mmHg}} \qquad \begin{array}{l} pK = 6.1 \\ \alpha = 0.03 \end{array}$$

$$\text{or} \quad [H^+] = K \frac{[CO_2]}{[HCO_3^-]} \quad K = 0.8 \times 10^{-6}\,M \qquad \begin{array}{l} \text{(or another} \\ \text{correct equivalent)} \end{array}$$

b) An excess of unwanted non-respiratory acid in the body.

c) Diabetes mellitus: defective carbohydrate metabolism and consequent metabolism of fats to yield ketone bodies (aceto-acetic and beta-hydroxy-butyric acids). These are relatively strong acids.

d) See Figures 2.2 and 3.4.

e) i) The initial addition of acid. $H^+ + HCO_3^- \rightarrow H_2O + CO_2$. CO_2 excreted by the lungs. Hence the fall in plasma $[HCO_3^-]$ with initially no change in P_{CO_2}. Rise in $[H^+]$, negative base excess indicating move away from the normal blood line. One might include an obliquity of arrow A downwards and to the right to represent initial release of CO_2 before CO_2 is removed by the lungs.

ii) The homeostatic corrections:

Respiratory: acidaemic drive to respiration via the peripheral arterial chemoreceptors gives hyperventilation. Acid–base status moves down the new blood line. Proportionally, P_{CO_2} changes more than $[HCO_3^-]$, so $[H^+]$ falls towards normal as a result; this is compensation for low $[H^+]$. Time course of the order of quarter of an hour, the time for CO_2 washout. Low P_{CO_2}, very low plasma $[HCO_3^-]$, negative base excess, partial restoration of $[H^+]$.

Renal: retention of bicarbonate (all filtered bicarbonate is retrieved, new bicarbonate is released into extracellular fluid), blood chemistry returns towards normal. Time scale: starts immediately but takes some days to be fully effective. Retention of bicarbonate alleviates the acidaemic respiratory drive and hence the obliquity of arrow BC in Fig. 3.4. All parameters restored towards normal. Final status depends on origin of acidosis; if due to a single dose of acid, then restoration is complete. If fixed acid is continually being released into the body as in diabetes, then compensation is partial. If onset of acid accumulation is gradual, then the phases cannot be distinguished.

Role of bone: buffering of acid, long time course because mineral in bone is inaccessible to the blood. Osteoporosis in chronic metabolic acidosis.

Learning objectives in acid–base physiology

CHAPTER 1

Water

1.1. Describe and explain properties of water which are important in physiological mechanisms.
1.2. State how much water there is in the body. Explain why we differentiate between "intracellular' and "extracellular' fluid; give the magnitudes of each.
1.3. Write the reaction for the dissociation of water and indicate the general significance of the fact that water dissociates.
1.4. Define 'acid', 'base' 'alkali'. Write the reaction for the dissociation of an acid; describe what is meant by 'strong' and 'weak' acids.
1.5. State and explain the law of mass action.
1.6. Define and write an equation for the 'ionic product of water'; give its value.

pH and buffers

1.7. Define pH.
1.8. Calculate the approximate pH of a given concentration (e.g. M, M/10, M/100) of strong acid or alkali.
1.9. Explain what is meant by buffering. Write down the chemical reactions occurring when a weak acid and its salt are added to an aqueous solution. Identify the buffer acid and the buffer base.
1.10. Explain why a mixture of a weak acid and its conjugate base acts as a buffer.
1.11. Derive the Henderson-Hasselbalch equation.
1.12. Define the pK of a weak acid or base.

1.13. Define and explain 'buffer value', describe how, for a particular buffer, this varies with pH. Explain why a buffer pair has its greatest buffer value at a pH which equals its pK.

CO_2–bicarbonate system

1.14. Derive the expressions $pH = pK + \log[HCO_3{}^-]/[CO_2]$. State that the pK in this instance is 6.1 when $[HCO_3{}^-]$ and $[CO_2]$ are measured in M.

1.15. State the relationship between concentration of a gas in dilute solution and its partial pressure; define 'solubility coefficient'.

1.16. State that $[CO_2]$ in mM equals $\mathbf{a} \times P_{CO_2}$, P_{CO_2} in mmHg and $\mathbf{a} = 0.03$ (the solubility coefficient for CO_2 in water).

1.17. Explain that the CO_2 and $HCO_3{}^-$ buffer pair is a poor chemical buffer because its pK is more than one pH unit away from the pH of the body. Explain that the buffer pair is physiologically important because the body can manipulate the concentrations of both the buffer acid (via the lungs) and the buffer base (via lungs and kidney).

CHAPTER 2

The range of extracellular $[H^+]$

2.1. Give a typical value for the extracellular pH and the limits normally compatible with survival. Give corresponding values for $[H^+]$. Compare the physiological range with that for sodium and chloride ions; explain any difference.

2.2. Explain the consequences of wide variations of extracellular $[H^+]$.

2.3. Indicate why hyperventilation may result in hypocalcaemic tetany; include the importance of extracellular $[H^+]$ variation.

2.4. Describe which tissues provide buffering power.

Graphic representation of acid–base status

2.5. Describe in outline how the concentrations of H^+, CO_2 and $HCO_3{}^-$ of a solution are estimated.

2.6. On a graph of P_{CO_2} (abscissa) and $[HCO_3{}^-]$ (ordinate), draw lines joining points of equal pH.

2.7. On the graph of 2.5, show and explain the effects of changes in P_{CO_2} on $[HCO_3{}^-]$ for:
   (a) distilled water;
   (b) sodium bicarbonate solution;
   (c) sodium bicarbonate solution plus a non bicarbonate buffer.

2.8. On a graph of $P_{CO_2} - [HCO_3{}^-]$, draw the relationships for normal plasma and for normal blood *in vitro*. The graph should show typical numerical values for P_{CO_2} and $[HCO_3{}^-]$ in normal arterial blood.

CHAPTER 3

3.1. Explain that CO_2 can be regarded as **volatile**, since it can be retained or removed by the lung. $HCO_3{}^-$ can be retained or removed by the kidney. Other acids and alkalis are non-volatile or fixed.

Disturbances of acid–base balance

3.2. Define – acidaemia and acidosis, alkalaemia and alkalosis.

Respiratory disturbances

3.3. On the graph of $P_{CO_2}{}^- [HCO_3{}^-]$, indicate the immediate effects of hypo-ventilation and hyperventilation and deduce the effects on $[HCO_3{}^-]$ and $[H^+]$. Also on the graph, indicate and explain the delayed effects, mediated by the kidney, occurring if the primary respiratory disorder persists. Hence define uncompensated and compensated respiratory acidosis and alkalosis.
3.4. Define partial, complete and over-compensation.
3.5. Explain that the renal compensation of altered $[H^+]$ involves alterations in blood biochemistry to restore the blood pH; compensation does not mean the restoration of normal blood biochemistry.

Metabolic disturbances

3.6. Define and explain the meaning of 'metabolic disturbances of acid–base physiology'. Give examples.
3.7. With the help of the normal blood line, describe and explain:
   (a) the immediate effects on blood biochemistry of a bolus injection of a strong fixed acid or alkali;
   (b) the respiratory compensation in each case;
   (c) the renal compensation in each case.
   Give the approximate time courses of these compensatory processes.

Potassium

3.8. State the relationship between extracellular concentrations of $[H^+]$ and $[K^+]$.

3.9. State and explain factors which contribute to this relationship.

3.10. State the principal adverse effects of hyperkalaemia and hypokalaemia.

Acid–base status and persistent vomiting of gastric contents

3.11. Describe the chemicals which are lost from the body in persistent vomiting of gastric contents.

3.12. Explain the acid–base disturbance produced by persistent vomiting of gastric contents.

3.13. Explain the renal mechanisms called into play to compensate for the fluid loss; indicate why the compensation for fluid loss exacerbates the acid–base disorder.

3.14. Describe the movements of potassium that occur in a patient suffering from chronic vomiting.

3.15. From the above, indicate how it is that some patients arrive at the situation of having extracellular alkalosis and intracellular acidosis.

3.16. Explain how the cautious intravenous infusion of physiological saline is likely to improve the patient's condition.

CHAPTER 4

Assessment of acid–base status

4.1. Define the **standard bicarbonate** for blood and explain how it is measured; give a typical normal value. Show this point on the 'normal blood line'.

4.2. State that a change in standard bicarbonate is indicative of a metabolic component in an acid–base disorder. Explain this on the graph of the blood buffer line and explain why the magnitude of the change in standard bicarbonate is an underestimate of the magnitude of the metabolic disorders.

4.3. Explain that the blood buffer base consists mainly of HCO_3^- and Protein-(Pr^-). Define the **buffer base** as ($[HCO_3^-] + [Pr^-]$).

4.4. Explain that alteration of the P_{CO_2} does not significantly alter the magnitude of the buffer base. Explain that addition of a given amount of strong acid or alkali to a blood sample results in a quantitatively equal change in buffer base.

4.5. From the preceding, and in relation to the graph of the 'normal blood line', define the terms **base excess**. Define 'base deficit' and describe how it can be regarded as a 'negative base excess'.

4.6. Explain how base excess can be directly determined by a 'back-titration' of 'titratable' acid or alkali. Explain why a P_{CO_2} of 40 mmHg is used and a pH of 7.4 is the end point of the titration.

4.7. Explain that the acid–base status can be deduced from the blood biochemistry in cases of uncompensated respiratory or metabolic disorder.
4.8. Demonstrate examples of compensated disorders (e.g. compensated respiratory acidosis and compensated metabolic alkalosis) in which the same abnormality of blood biochemistry may be reached by different routes. Hence explain the need, in such cases, to seek each cause and evaluate its effect.

The Siggaard-Andersen nomogram

4.9. Explain the advantage of using $\log(P_{CO_2})$ as a function of pH for constructing a nomogram to measure the parameters of acid–base status.
4.10. State and explain the plots given by:
  (a) a solution with; no buffering power;
  (b) a perfect buffer;
  (c) plasma;
  (d) blood.
4.11. Explain the construction of the scale for reading off base excess.
4.12. Explain the analyses needed on a blood sample in order to be able to plot a line on the nomogram and how to use the nomogram to read off standard bicarbonate, base excess, total buffer base.
4.13. Define **normal buffer base**, describe how to calculate it and explain its value.
4.14. Indicate how variations in the haemoglobin concentration influence the different parameters of acid–base status and how this is accommodated in interpretation of results.

CHAPTER 5

Sources and buffering of acid and alkali

5.1. Describe the sources of acid and alkali in the body, giving approximate values.
5.2. Describe and explain the contributions of different buffer systems to respiratory and metabolic disorders of metabolism.

Composition of plasma

5.3. Give a balance sheet for the concentrations of electrolytes in plasma.
5.4. Give the quantitative contribution of proteins to (a) electrical neutrality and (b) the colloid osmotic pressure of plasma.

Plasma and erythrocytes in whole blood

5.5. Explain what is meant by separated plasma and true plasma.

5.6. Does the measurement of pH on a sample of whole blood give a value representing an average for erythrocytes and plasma, for erythrocytes or for the plasma component? Explain your answer. Explain how you would calculate the $[HCO_3{}^-]$ from measurements of pH and P_{CO_2}. (Why does the result apply to the plasma component when the measurements are made on whole blood?).

Whole-body acid–base status

5.7. Compare the buffer lines for blood *in vitro* and *in vivo*. Explain the difference. Include the effects of
(a) the interstitial fluid;
(b) the buffering contributed by the cells of the body.

5.8. Explain the dangers in assuming that the acid–base status measured from a sample of arterial blood is quantitatively the same as that occurring in the body as a whole.

Intracellular pH

5.9. Give a typical value for intracellular pH.

5.10. Indicate and explain the changes in intracellular pH consequent on extracellular changes in respiratory and metabolic disturbances.

5.11. Describe in outline the regulation of intracellular pH and how this interacts with the regulation of extracellular pH.

CHAPTER 6

Carriage by blood of O_2 and CO_2

6.1. Draw the oxygen and carbon dioxide dissociation curves blood. For each, indicate and explain the significance of a change in partial pressure of the other gas.

6.2. Explain why the curves for carbon dioxide are close to, but not identical with, the $P_{CO_2} - [HCO_3{}^-]$ curve for blood.

6.3. Show and explain the 'physiological' oxygen and carbon dioxide dissociation curves for blood passing through systemic or pulmonary capillaries. Indicate and give the coordinates of points representing arterial and mixed venous blood.

6.4. Describe and explain the movements of molecules and ions through the red cell membrane during the passage of blood through
(a) the systemic capillaries;

(b) the pulmonary capillaries.

Include an explanation for the term 'chloride shift' and describe the role of carbonic anhydrase.

6.5. Describe and explain the changes in pH and erythrocyte volume as blood traverses the capillaries.

Asphyxia and hypoventilation

6.6. Describe, explain and indicate the time course of the arterial P_{O_2} and P_{CO_2} with cessation of ventilation (plastic bag over the head). Describe in outline the sequence of effects on the patient, with an indication of time scale.

6.7. Describe, explain and indicate the time course of the arterial P_{O_2} and P_{CO_2} with the sudden onset of hypoventilation (as in a severe attack of asthma).

Ventilation

6.8. Explain the control of ventilation in a healthy person at sea level breathing air.

6.9. Describe and explain the normal composition of alveolar gas and the way in which this composition varies in hypoventilation and in hyperventilation.

CHAPTER 7

Renal mechanisms in acid–base balance

7.1. Outline the need for renal defence of pH in:
   (a) a healthy human or carnivorous or mixed diet;
   (b) a healthy strict vegetarian.

7.2. Describe and explain the mechanism for the renal tubular excretion of H^+, including the role of carbonic anhydrase.

7.3. Describe and explain the mechanism of renal tubular retrieval of filtered bicarbonate.

7.4. Describe and explain the change of pH along the tubule in a person excreting acid urine.

7.5. Describe how the kidney can excrete large amounts of acid or alkali without producing urine whose pH varies outside the limits of 4.5 to 7.8. Include the role of phosphates and of ammonium.

7.6. Define 'titratable acid' and 'total acid'; indicate how they are measured.

7.7. Describe in outline the effects on acid–base homeostasis of a) renal failure and b) diuretic therapy.

References

Babkin, B. P. (1950) *Secretory mechanisms of the digestive glands*, 2nd edn, Hoeber, New York and London.

Bates, R. G. (1966) Acids, bases and buffers, *Annals of New York Academy of Sciences*, **133**, 25ff.

Barcroft, J. (1938) *Features in the architecture of physiological function*, Cambridge University Press, Cambridge.

Burton, R. F. (1992) The roles of intracellular buffers and bone mineral in the regulation of acid-base balance in mammals. *J. Comp. Biochem. Physiol.*, **102A**, 425–32.

Cohen, J. J. and Kassirer, J. P. (1982) Acid–base, Little, Brown and Co., Boston.

Davenport, H. W. (1974) *The ABC of acid–base chemistry*, 6th edn, University of Chicago Press, Chicago.

Davson, H. and Eggleton, M. G. (eds.) (1962) *Starling and Lovatt Evans Principles of human physiology*, 13th edn, Churchill, London.

Fleck, A. and Ledingham, I. McA. (1988) Fluid and electrolyte balance, in *Jamieson and Kay's Textbook of surgical physiology*, 4th edn, (eds) I. McA. Ledingham and C. MacKay, Churchill Livingstone, Edinburgh.

Friedman, M. H. F. (ed.) (1975) *Functions of the stomach and intestine*, University Park Press, Baltimore.

Guggenheim, E. A. (1957) *Thermodynamics: an advanced treatment for chemists and physicists.*, North Holland Publishing Company, Amsterdam.

Haldane, J. B. S. (1921) Experiments on the regulation of the blood's alkalinity, *J. Physiol.*, **55**, 26ff.

Heisler, N., Forcht, G., Ultsh, G. R. and Anderson, J. F. (1982) Acid–base regulation in response to environmental hypercapnia in two aquatic salamanders, *Respir. Physiol.*, **49**, 141–158.

Pitts, R. F. (1948) Renal excretion of acid. *Fed. Proc.*, **7**, 418–26.

Pitts, R. F. (1974) *Physiology of the kidney and body fluids* (3rd edn.), Year Book, Chicago.

Robinson, J. R. (1962) *Acid–base regulation*, Blackwell Scientific Publications, Oxford.

Siggaard-Anderson, O. (1963) Blood acid–base alignment nomogram, *Scand. J. Clin. Lab. Invest.*, **15**, 211ff.

Singer, R. B. and Hastings, A. B. (1948) An improved method for the estimation of disturbances of the acid–base balance of human blood, *Medicine*, **27**, 223–42.

Thomas, R. (1984) Experimental displacement of intracellular pH and the mechanism of its subsequent recovery, *J. Physiol. Lond.*, **354**, 3P-22P.

Valtin, H. (1983) *Renal function.*, 2nd edn., Little, Brown and Co., Boston.

Van Slyke, D. D. (1922) On the measurement of buffer values and on the relationship of buffer value to the dissociation constant of the buffer and the concentration and reaction of the buffer solution. *J. Biol. Chem.*, **52**, 525–70.

Van Slyke, D. D., Phillips, R. A. Hamilton, P. B. *et al.* (1943) Glutamine as source material of urinary ammonia, *J. Biol. Chem.*, **150**, 481–82.

Ypersele de Strihou, C. van and Frans, A. (1970) The pattern of respiratory compensation in chronic uraemic acidosis, *Nephron*, **7**, 37ff.

Glossary

Acid A proton donor. A chemical which raises the hydrogen ion concentration. An **acid** solution is one in which the concentration of hydrogen ions exceeds that of the hydroxide ions.

Acidaemia A condition in which the pH of the blood is below normal.

Acidosis A condition in which there is an excess of acid in the body.

Acute Sudden in onset, the opposite of chronic.

Alkalaemia A condition in which the pH of the blood is above normal.

Alkali A hydroxide ion donor. An **alkaline** solution is one in which the concentration of hydroxide ions exceeds that of the hydrogen ions.

Alkali reserve An obsolete term.

Alkalosis A condition in which there is an excess of alkali in the body.

Avogadro's law The volume occupied by one gram-molecule of a gas is the same for all gases under the same conditions of temperature and pressure. One mole of a gas in a volume of 22.4 litres exerts a pressure of one atmosphere (760 mmHg) at 0°C.

Base A proton acceptor.

Base excess (BE) This is the amount of strong acid (in mM) which must be added to one litre of oxygenated whole blood, equilibrated with carbon dioxide at a partial pressure of 40 mmHg and at 37°C., in order to bring the pH to 7.4.

Buffer Buffers are substances which by their presence in solution increase the amount of acid or alkali that must be added to cause unit change in pH.

Bolus A bolus injection means an injection given over a short period of time; this contrasts with an infusion, which is an injection over a long period of time.

Buffer base (BB) The sum of the concentrations of reaction-dependent anions, quantitatively the most important being bicarbonate and protein anions.

Buffer value The amount of strong acid or alkali (in moles per litre) producing a change of one pH unit.

Carbonic anhydrase An enzyme which catalyses the chemical reaction $CO_2 + H_2O = H^+ + HCO_3^-$.

Chronic Developing slowly, of long standing, as in 'chronic bronchitis'.

Colligative A colligative property of a solute in a dilute solution is an effect whose magnitude depends only on the concentration of the solute particles (number of particles per unit volume of solution) and not on features such as size, shape or chemical composition. Examples are osmotic pressure, depression of freezing point, elevation of boiling point and the vapour pressure of the solution. If the value of any one of these is known for a particular solution, then the values of the others may readily be calculated.

Compensation In acid base physiology, this means a homeostatic mechanism which has the effect of returning the extracellular hydrogen ion concentration towards normal when it has deviated.

Conjugate (L. conjugatus – joined together) Paired. The conjugate base of an acid is the base which, when conjugated (joined) to a proton, yields that acid.

Dalton's law of partial pressures When two or more gases are mixed, the total pressure is equal to the sum of the partial pressures of the constituents.

Electrolyte An electrolyte is a chemical which carries electricity if a current is flowing in solution. It is charged; examples are sodium and bicarbonate ions.

Equivalent Is a measurement of quantity of charge; equiv. only applies to ions; numerically it is given by moles multiplied by valency.

Fixed This adjective has two different meanings in acid–base physiology.

Fixed acid Is synonymous with non-respiratory acid. **Fixed anion** is an anion whose concentration is unaffected by a change in pH. Chloride is a fixed anion whilst bicarbonate is not.

Henry's law In a dilute solution, the concentration of the gas in physical solution is directly proportional to the partial pressure of the gas. The constant of proportionality is the **solubility coefficient** of the gas.

Hydrogen ion A hydrogen ion is a proton, i.e. a hydrogen atom from which the electron has been removed.

Hyperventilation Overbreathing; a condition in which the alveolar P_{CO_2} is below normal.

Hypoventilation A condition in which the alveolar P_{CO_2} is above normal in a person breathing air; inadequate ventilation.

Iatrogenic Caused by medical treatment such as drugs intended to cure disease.

Infusion Intravenous infusion. An injection given over a long period of time.

Ion A particle carrying an electric charge.

Ketone bodies Strong acids produced by the abnormal metabolism of fats in diabetes mellitus.

Law of mass action The velocity of a chemical reaction at constant temperature is proportional to the product of the concentrations of the reacting substances.

mEquiv A measurement of quantity of charge expressed in units of milli-equivalents (there are 1000 milliequivalents in one Equiv.).

Metabolic disorder of acid–base physiology A disorder of acid–base balance not caused by a respiratory disorder. An example is the excess of acid which occurs in uncontrolled diabetes mellitus.

Mitochondrion (plural a) An intracellular organelle on which are laid out the enzymes subserving aerobic metabolism.

Mixed disorder of acid–base physiology This expression is used with two different meanings. Some authors use it to indicate the co-existence of two primary abnormalities, e.g. respiratory acidosis and metabolic acidosis. Others use it as a synonym for 'compensated', when a primary disorder is accompanied by a super-added physiological response, e.g. compensated respiratory acidosis, when the primary disorder is acidosis and the renal compensation adds a physiological metabolic alkalosis.

Molar Is a measurement of concentration in units of moles per litre.

Mole Is a measurement of quantity of a chemical; it is the molecular weight in grams.

M Is a measurement of concentration in units of moles per litre, or molar.

mM Represents millimoles per litre (there are 1000 millimoles in one mole).

Naturetic A naturetic agent is one which promotes the loss of sodium in the urine.

Neutral A **neutral** solution is one in which the concentrations of H^+ and OH^- are equal.

nM Represents concentration in units of nanomoles per litre (there are 1000,000,000 nanomoles, or 10^9 nanomoles, in one mole.

Normal buffer base (NBB) The concentration of buffer base when any changes in the magnitude of the buffer base contributed by disturbance of acid–base status have been removed. $NBB = BB - BE$.

Paralytic ileus A condition of paralysis of the smooth muscle of the gut. Peristalsis ceases, alimentation ceases and frequently large volumes of fluid accumulate in the bowel. This acts as a space-occupying lesion and may interfere with activity of other viscera.

Partial pressure The pressure which any one gas exerts, whether it is alone or mixed with other gases, is the 'partial pressure' of the gas.

pH The negative logarithm of the hydrogen ion concentration. $pH = -\log[H^+]$, where the concentration $[H^+]$ is in moles per litre.

pK The negative logarithm of the dissociation constant of an acid.

Principle of electroneutrality In a solution, the number of positive charges is equal to the number of negative charges.

Proton Atoms are made up of three subatomic particles, protons, neutrons and electrons. A hydrogen atom consists of one proton and one electron. A proton can exist in a free form as a hydrogen atom from which the electron has been removed. It is alternatively called a **hydrogen ion**.

Reaction In acid–base physiology, 'reaction' is similar in meaning to 'pH'. The 'reaction of a solution' means the 'pH of a solution'. An anion whose concentration is reaction-dependent is one whose concentration changes with addition of acid or alkali, e.g. bicarbonate.

Respiratory disorder of acid–base physiology A disorder of acid–base balance primarily due to a disorder of respiratory function, hypoventilation in which the concentration of carbon dioxide in the blood is increased or hyperventilation in which the opposite change occurs.

Respiratory exchange ratio Is the CO_2 output divided by the oxygen uptake. At times when the excretion of CO_2 matches its production, the respiratory exchange ration equals the RQ. If excretion exceeds production, the respiratory exchange ratio exceeds the RQ In the early phases of hyperventilation, the respiratory exchange ratio may approach 2.

Respiratory quotient (RQ) is the volume of carbon dioxide produced divided by the volume of oxygen consumed in the tissues in a given period of time. The value of the RQ depends on the nature of the metabolic energy source. For carbohydrates, the RQ is 1, for fats it is 0.7 and for proteins it is 0.8. A healthy subject on an ordinary mixed diet of carbohydrate, fat and protein has an RQ of typically 0.85.

Solubility coefficient or constant The volume of a gas which dissolves in physical solution per unit increase in partial pressure of the gas.

Standard bicarbonate This is the concentration of bicarbonate in the plasma of oxygenated whole blood after equilibration with carbon dioxide at a partial pressure of 40 mmHg at 37°C.

Systemic The systemic circulation is the blood circulation excluding the pulmonary circulation. The systemic arteries are all the arteries other than pulmonary arteries and similarly for capillaries and veins.

Titratable acid The 'titratable acid' of the urine is measured by titrating the urine produced in a given period of time back to the pH of the blood. This gives the number of hydrogen ions free in the urine and buffered by the urinary buffers such as phosphate. It does not measure the hydrogen ions buffered by ammonia, because the ammonia is produced by the renal tubular cells themselves.

Titration curve A plot of the pH of a solution as a function of the amount of strong acid or strong alkali added to it.

Total acid excretion The 'total acid excretion' of the urine is the quantity of ammonium ions found in the urine plus the titratable acid.

Total buffer base See buffer base.

Units and abbreviations

The information given here is intended to help the reader who is confused by differences in nomenclature and abbreviations when comparing this book with other accounts.

Concentrations: In this book, concentrations have in most cases been expressed in moles per litre. The abbreviation used for this has been **M**. Other methods of representing this same unit are: mole/litre and mole l^{-1}.

Units of pressure: In this book, pressure has been measured in units of mmHg. This is the unit most commonly encountered in acid–base physiology but, in respiratory physiology, there is a move to change to the use of **kilopascals**, abbreviated kPa. A pascal is one newton per square metre; one newton is the force which, when applied to a mass of one kilogram, produces an acceleration of one metre per second per second. The conversion between mmHg and kPa is:

$$750\,mmHg = 100\,kPa$$

or, equivalently

$$1\,kPa = 7.5\,mmHg.$$

At sea level, the atmospheric pressure is typically 760 mmHg. It is a fortuitous circumstance that one atmosphere is approximately 100 kPa.

Index

Acetoacetic acid   81
Acid   3
  definition of   3
  excretion of   128–9
  strong   3
  titratable   132
  weak   3, 89
Acidaemia   50
Acid–base   143, 157–8
  balance   106–8
  disorders   119–20
    treatment   94–6
  homeostasis   96–8
  status   135, 172, 172–3
    indicators of   60
    specification of   67–74
Acidaemia
  an advantage of   46
  definition of   38–40
Acids and bases in the body   82
Acidosis
  definition of   38–40
  uncompensated metabolic   44
Activity   18
  definition   18
  of a chemical   18–19
Acute respiratory acidosis   35
Acute hypoventilation, see Respiratory
    acidosis
Adipose tissue   1
Aeration of the blood in the pulmonary
    capillaries   33
Aerobic metabolism   105
Aldosterone   132

Alkalaemia   50
  definition of   38–40
Alkali
  definition of   3
  reserve   179
Alkaline   2
  tide   52
Alkalosis
  and hypokalaemia   41
  definition of   38–40
Alveolar gas   114–15
Alveolar membrane   115
Ammonia   129–31
  formation in the body   130
  production in kidney   130
  and renal acid–base physiology
    130–1
Ammonium ions   130
Ammonium trap   131
Anaemia   24, 65
Anaerobic metabolism   42, 105
Aqueous solutions   18
Arterial plasma   21
  pH range   21
Asphyxia, death due to   112
Asthma   33
Athlete   81
Atmospheric air, composition of   114
Avogadro's law   17

Base   3
  deficit   63
  definition of   3
Base excess (BE)   62–3

Base excess (BE)—*contd.*
  as an empirical measurement 93
  negative 63
BB, *see* Buffer base
BE, *see* Base excess
Bicarbonate 14, 16
  concentration 16
  extracellular concentration 95
  filtered 125
  generation, renal 16
  infusion 95
  intracellular concentration 95
  renal excretion 54
  movement across erythrocyte
    membrane 92
  renal control 14
  renal handling 125
  retention by the kidneys 38
  space 95
  standard 60
  transfer across cell membranes 92, 96
Blood 31
  buffers 35, 41–2, 87–8
    bicarbonate and non-bicarbonate
      buffers 42
  consists of two phases 90
  in the body 93–4
  plasma 31
Body fluids 7, 83
Bohr effect 104–6
Bone 24, 44
  as a buffer 44
Brain 97
Bronchioles, narrowing of the 33
Bronchoconstriction 33
Buffer 7
  blood 31, 41–2, 72
  bone as a 44
  non-bicarbonate 42
  pair, pK of 11
  perfect 30
  power 12
  real solutions 30–1
  protein 42
  systems of the body 23
  value 12, 19
Buffer acid 9
  non-bicarbonate 42

Buffer base (BB), synonymous with
  'total buffer base', 9, 63–5
  measurement of 79
  non-bicarbonate 42
  normal 74–7
  of plasma and of whole blood 86
Buffering
  action of the body 24
  by bone 24
  efficacy 11
  of an acid load 82
  of haemoglobin 65
  of strong acid 10
  of strong alkali 11
  power of the blood 65
Buffers 169
  blood 35, 41–2
Buffer pair 8
Buffer value 12, 19
  definition 12
Burns 81

Capillary membrane 92
Carbon dioxide 14, 26, 99, 111–14,
  117–19
  accumulation of 33
  and oxygen: the sea-saw 116–17
  and respiratory acidosis 111
  carriage in blood 109
  chemistry 32
  control of partial pressure in
    blood 109
  dissociation curve of the blood 106–8
    'physiological' curve 109
  hydration of 32
  hydroxylation of 32
  metabolic production 115
  released by metabolism 32
  regulation of PCO_2 115
Carbonate 32
Carbonic
  acid 89, 99
  anhydrase 123
    inhibitors 132
Cardiac disease 132
Catabolism 81
Cell membrane
  permeability to blood gases 47

permeability to ions   47
Central chemoreceptors   120
Charge on one ion of protein buffer
    base   87
Chemical, 'activity' of   18
Chemoreceptors   120
Chloride
    ions   22
    loss with vomiting of gastric
        contents   55
    plasma concentration   55
'Chloride shift'   91, 92
Chronic, definition   36
Circulatory collapse   55
Citric acid   95
Closed system   11
CO_2–bicarbonate system   14, 170
Colligative   180
Colloid osmotic pressure   87
Compartments, fluid   1
Compensation   35, 44, 180
    degrees of   36–7, 45
    is physiological buffering   36
    of respiratory acidosis   36
    renal   46
    time scale   37, 45
Compensatory mechanism
    renal   36
    respiratory   44
Complex protein molecules   22
Composition of atmospheric air   114
Composition of plasma   83
Concentration of a constituent of a
    gas mixture   17–18
Constancy of the hydrogen ion
    concentration   97
Conductivity measurements   7
Conjugate   180
    base   3, 91
Control of ventilation   120–1
Curve, titration   7

Dalton's law of partial pressures   17, 180
Davenport plot   78
Dead space   114
Death due to asphyxia   112
Defence of hydrogen ion
    concentration   23

Defence of the extracellular fluid
    volume   56
Degrees of compensation   45
Depression of respiratory centres   33
Diabetes mellitus   42
Dilute aqueous solution   19
Dissociation
    constant   9
    of salts   83
    of water   2, 4–6
Distilled water   27–9
Disturbances of hydrogen ion
    concentration   23
Duodenal ulcer   47
Diuretics   132
Diuretic therapy   132–3
Donor
    hydroxide ion
    proton   3
Drug abuse   33
Duodenal contents   53

Effective concentration   19
Electrical balance   92
Electrical potential   49
Electrode, pH   7
Electrolyte   180
Electron   2
Electroneutrality   49, 85, 125
    principle of   85
Enzymes   7, 22
    ionization of   23
Equilibrium   4
    constant(s)   4, 89
    definition   4
Equivalent   83, 180
    definition   84
Erythrocyte, buffering by   31, 72
    a non-nucleated cell   97
Evaporation   1
Excretion
    of acid   128–9
    of bicarbonate   16, 125
Exercise   42, 81, 115
Extracellular
    electrolytes   22
    fluid(s)   1, 22
    volume   22

Fever   81
Fraction of gas by volume   17
Filtered bicarbonate   125, 126
Filtered load of chloride ions   56
Fixed   180
Fluid compartments   1

Gastric acid   52
Gastric alkalosis   56, 96
Gastric juice   52
  composition of   52–3
Gastrin   54
Glomerular filtrate   123
Glutaminase   130, 131
Graphic representation   25–7
  of acid-base status   24, 170–1

Haemoglobin   42, 65
  buffering by   31, 72
  concentration   103
  binding sites   103
Haldane effect   110
Harmonious action of enzymes   97
Henderson-Hasselbalch
    equation   9–10, 88–9
Henry's law   18, 26, 180
Hepatic disease   132
Homeostasis   96
Homeostatis   97
Hydrochloric acid   43, 91
  infusion of   43
Hydrogen atom   2
Hydrogen ions   2, 96, 180
  active pumping   98
  comparison of, with other
    electrolytes   22
  concentration   5
    defence of   23
    effects of disturbances of   23
    enzymes, and   22
    extracellular   21
    intracellular   96
  inter-relationships with potassium
    ions   51
  secretion along nephron   128
  transfer across the cell
    membrane   96–8
Hydroxide ion   2

donor   3
Hypocalcaemic tetany   23
Hypokalaemia   23, 41, 50
Hypovolaemia   55
Hyperkalaemia   50
Hyperventilation   117–19, 180
Hypoventilation   113, 180
  acute   33
  defined   33

Induction   131
Infusion   180
  of hydrochloric acid   43–4
  of strong acid   43
Inhaling a foreign object   33
Inspired air   114–15
Internal environment   96, 97
Intracellular acidosis accompanying
    the extracellular
    alkalosis   57
Interstitial fluid   92
Intracellular fluid   1
Ion   180
Ionic
  product   5
    of water   5, 18
  variations in composition of   22
  strength   18
Ionization of enzymes   23
Irritation of the small bowel   53

Kidney   16–17
Ketone bodies   180

Lactic acid   42
Latent heat of evaporation of water   1
Law of mass action   3, 4, 24, 180
  definition   4

Mass action   4, 35, 129
Making a buffer solution   8
Mass action   24
Measurement of acid-base status   59–60
Measurement of pH   7, 18
Measurement of quantities and
    concentrations   xiii
Mechanical ventilation   23
Membranes of animal cells   83

Metabolic
  acidosis 42, 94, 121
    causes of 58
  alkalosis 47, 142, 157
    causes of 47
  disorder, definition of 41
  disorder of acid–base physiology 181
Metabolism, abnormal 41
mEquiv xiii, 181
Milieu interieur 95
Mitochondrion (pl. mitochondria) 181
Mixed disorder of acid–base
    physiology 181
Mixed expired gas 114
mM xiii, 181
Molar xiii, 181
Molarity xiii
Mole xiii, 181

Naturetic 132, 181
NBB, see Normal buffer base
Neutral solution 181
Nephron 128
New blood line 37–8
Nitrogen 114
nM xiii, 181
Normal blood line 35
Normal buffer base (NBB) 74–7, 181

Obstruction to the air passages 33
Open system 14
Osmotic gradient 83
Osmotic pressure 83
Oxygen 114
  carriage 105
  carrying capacity 101
  cascade 106
  dissociation curve of the
    blood 101–3, 121
  'physiological' curve 105
  off-loading 111
  partial pressure 101

Paralytic ileus 181
Partial pressure 17, 181
  of gases 17, 115
    in alveolar gas 115
    in systemic arterial blood 115
Pasteur point 105

Peptides 3
Perfect buffer 30
Peripheral chemoreceptors 120
pH 123, 169
  definition of 5, 18, 181
  electrode 7
  of extracellular fluid 14
  measurement of 7, 18
  meter 7
  range of 5–6
  scale of 5
Phosphate 129
Phosphate as a buffer of urine 129
Phosphofructokinase 96
pH range, arterial plasma 21
pK 181
  definition 9
  use of the prime 99
Plasma
  composition of 83, 85–6
  proteins 42
  separated 90
  true 90
Plateau 103
  of oxygen dissociation curve 103
Potassium 22, 48, 143, 158, 171–2
  adverse effects of changes 51–2
  balance 56
  concentration in arterial plasma 22
  content of the body 57
  extracellular concentration 48
  filtration 58
  loss with vomiting 55
  reabsorption 58
  renal handling
  secretion 58
Potassium and acid–base
    physiology 48–9
Primitive organisms 97
Principle of electroneutrality, 181
  definition 83
Production of acid in the body 82
Properties of gases 17–18
Proteins 3
Protein buffer 42
Proton 2, 181
  acceptor 3
  donor 3

Pulmonary capillaries
  aeration in the blood in the   33
Pulmonary disease   132
Pyloric obstruction   54

Range of plasma pH   21
Reaction   182
Real buffer solutions   30–1
Rebreathing   111
Red blood corpuscle, *see* Erythrocyte
Regulation of respiration   115
Renal
  compensation   24
  compensation for respiratory
    acidosis   57
  compensatory mechanisms   35
  disease   132
  excretion of bicarbonate   54
  failure   132
  handling
    of electrolytes   49–51
    of potassium   58
  interstitial fluid   124
  secretion of hydrogen ions   123–7
  tubular pumping of hydrogen
    ions   123
Respiration, control of   115
Respiratory
  acidosis   33, 111–14, 141, 156
    acute hypoventilation   33–6
    timescale   35
      renal compensation of   37
  alkalosis   33, 40–1
  compensation   23–4
    for metabolic acidosis   45
  centres, depression of   33
  disorder of acid–base
    physiology   33–58, 182
  exchange ratio   119 (definition),   182
  quotient (RQ)   110, 182
RQ, *see* Respiratory quotient

Salamander   98
Scale, pH   5
Sea   97
'Sea within us'   97
See-saw of carbon dioxide and

  oxygen   116–17
Separated plasma   90
Siggaard-Andersen nomogram   67, 163
  construction of   72
  use of   76
Sodium bicarbonate   47, 95
  solution   29
Sodium citrate   95
Sodium lactate   95
Sodium ions   22
Solubility coefficient or constant   18, 26,
  182
Solvent   1
Sorenson   18
Sources of acid or alkali   81
Specific heat of evaporation of water   1
Standard bicarbonate   60–2, 182
Starvation   81
Strong acids   3
Strong acid administered
  intravenously   43
Suffocation   111
Surgery   81
Sweat, evaporation of   1
Systemic blood   115–16, 182

The standard bicarbonate   60
Time scale   37
  of hyperventilation in metabolic
    acidosis   45
  renal compensation of respiratory
    acidosis   37
Titratable acid   132, 182
Titration curves   7, 182
  and buffers   7
Tonometer   101
Total
  acid excretion   132, 182
  buffer base   63–5, 182
    of plasma and of whole blood   86
Transfer of charge *see* Electroneutrality
Trap, ammonium   131
Trauma   81
Tubular reabsorption   16, 125

Uncompensated   35
  definition   35

metabolic acidosis   44
  respiratory   35
Urine pH-limits   123

Velocity coefficient   4
Ventilation, control of   120–1
Ventricular
  arrhythmia   23
  fibrillation   23
Volume of circulating blood   22
Vomiting   53
  effect on acid–base status   56
  of gastric contents   54–7, 142, 157, 172

Water   1, 169
  in the body   1
  ionic product of   5
  latent heat of evaporation of,   1
  specific heat of   1
  the substrate of life,   1
  vapour pressure   115
    dependence on temperature   115
    in alveolar gas   115
    in air   114
Weak acids   3
Whole-blood buffering of an acid
    load   82
Whole-body acid–base physiology   92